Kuwait: Prospect and Reality

KUWAIT
Prospect and Reality

by H. V. F. WINSTONE and ZAHRA FREETH

Distributed in the United States by
CRANE, RUSSAK & COMPANY, INC.
52 Vanderbilt Avenue
New York, New York 10017

CRANE, RUSSAK & COMPANY, INC
NEW YORK

First published in 1972

ISBN 0-8448-0020-1

Published in The United States by:
Crane, Russak & Company, Inc.
52 Vanderbilt Avenue
New York, N.Y. 10017

Library of Congress Catalog No. 72-80110

Printed in Great Britain

To Arab friends and friends of Arabia

1. The Amir of Kuwait, His Highness Shaikh Sabah al Salim al Sabah

Apologia

The idea of a general history of Kuwait was suggested by the chance circumstance of my being asked to write an account of the country's oil concession. In performing that task it became evident that almost every aspect of this fascinating corner of the Gulf had been dealt with in near isolation. The oil story, for example, had been covered by a legion of writers with reference to the companies involved, or to British and American commercial activities in the Middle East. Seldom, if ever, do these authors relate the advent of oil to Kuwait's own historical development. Similarly, official and semi-official documents produced for the Kuwait government are, understandably, inclined to concentrate on the post-oil achievements of the country to the exclusion of the past. The few attempts that have been made to bring past and present together have tended to be either sycophantic, and therefore suspect, or destructively critical. One cannot help wondering what the British or Americans would think, were an Arab writer to make a brief excursion to their countries and proceed, on the strength of hearsay evidence, to conjecture about the private lives of their citizens, question their political institutions and patronise their religious customs. Such has been the habit of some western writers let loose in the Arabian peninsula. Happily, the Arab endures the impertinence with good-natured fortitude.

There is a notable exception to the shortcomings of Kuwait's considerable literature. For several decades one British family has known the country intimately and recorded its fortunes with affection and felicity. The late Colonel H. R. P. Dickson, Political Agent, adviser to the country's ruling family and an important participant in the oil negotiations, described the people, the events and customs of old Kuwait with a wealth of detail. He knew and loved the Arabs of town and desert, pursuing their aspirations and interests with understanding. Without his carefully compiled and documented accounts of Kuwait before oil and material wealth came to change its face, it would be impossible to construct an objective history of one of the country's most crucial periods. Col Dickson's work in Kuwait was shared by his wife Violet. Since the late 1920s, when the Dicksons came to Kuwait, she has remained the most respected expatriate woman of the entire Gulf region, a source of information and wisdom to whom visitors of every nationality and purpose invariably beat a path. While her husband attended to his political and literary tasks, she recorded the fast-disappearing flora of the desert, and gathered impressions which were later to appear in the form of reminiscences embracing forty years of participation in the life of Kuwait.

It was my good fortune to meet Mrs Dickson and, through her, Zahra Freeth, her married daughter. Out of the latter meeting grew the notion of this joint enterprise, for though I had acquired a useful grounding in the oil story I was certainly not qualified by knowledge or association to write with familiarity of Kuwait or its people.

The structure of the book is mostly mine, the gist – or at any rate that part of it which relates directly to Kuwait – largely hers. Such weaknesses as there may be in geological, archaeological or economic reasoning must be attributed to me.

There is one other circumstance which convinced us of the need for a comprehensive portrait of Kuwait. Recent archaeological investigations along the Gulf coastline from Kuwait to the Trucial States have thrown an entirely new light on the extent and character of its past. Until a Danish team came to excavate in 1953, Kuwait's links with the distant past were tenuous, relying largely on deductions from the general history of this part of Arabia. Now it is known beyond reasonable doubt that Kuwait was part of one of the earliest civilisations known to man, pre-dating perhaps even the historic communities of the alluvial plains to the north.

In thus attempting to bring past and present within the compass of a single volume we have worked to a broad perspective. It is difficult to tell the story of any country without regard to what went on before and around. Kuwait is no exception.

Apart from the readiness of the Kuwait Government and its Embassy in London to supply statistical information, we have neither sought nor received official assistance. Similarly, in telling the story of oil in Kuwait within the framework of the country's social and political development, we received assistance from the Kuwait Oil Company; but we did not ask for or receive its sanction. Reference to the search for the lost civilisation of the Gulf leans heavily on Mr Geoffrey Bibby's magnificent book *Looking for Dilmun*. Inevitably, with so stretched a canvas we have had to rely a good deal on the scholarship and research of others. We hope that we have made amends by textual acknowledgement and in bibliography. Facsimiles and transcripts of Crown-copyright records are reproduced by permission of the Public Record Office, the India Office and the Controller of HM Stationery Office.

For permission to use photographs of the early exploration period in Kuwait we thank Mr and Mrs L. D. Scott, Mr David Curtis and the Kuwait Oil Company. Our thanks are also due to Mr A. N. Donaldson for permission to draw on his monograph "The History of the Kuwait Postal Service", Roman and Moira Buj for their advice and contributions to graphic illustration of the book, to Ronald Smith for his drawings of objects from Dilmun, and to Maureen Morgan who researched and typed. Finally, a note on Arabic usage. In transliteration we have in most cases bowed to popular preference. For example, we would have preferred the more exact *badu* and *badawin*, but we have settled for *bedu* and *beduin*. We stop short at the Franco-Arabic *bedouin*. In the spelling of *Kuwait, Shaikh* and other commonly used words we have maintained consistency even in quotations, thus taking slight liberty but, we hope, saving the reader much irritation.

H. V. F. W.
London 1971

Introduction

In simple translation, Kuwait is the Little Fort. To Europeans who skimmed the surface of Arabia for five hundred years or so it was a sandy littoral, a lip of the Gulf, south of Basra, which gave them no great need to pause in their tracks. They also called it Qurain, the Little Horn. Or Grane, or sometimes Grain.

Today, it is one of the richest nations on earth, engaged in a massive experiment in social welfare and democracy, maintaining on one level its traditional Arab character – stoic, proud, fatalist – while on the other it balances a modern economy, generated by a gush of oil of almost unimaginable abundance.

Perhaps the duality of Kuwait is best revealed by its visual contrasts. By dhows, often neglected, yet altogether more patently to do with the crafts of shipbuilding and the skills of seamanship than the gigantic oil tankers which nowadays lumber to and from its oil jetties. By the herdsmen who come and go as they have come and gone for several thousands of years, with their camels and sheep and goats, accepting without question the oil wells and electrical transformers which they pass on their way. By bedu encampments whose goathair tents are juxtaposed to desert villages with bristling television aerials. By a vast expanse of modern building that is almost alarming in its manifestations of architectural fallibility, dwarfing the few remaining mud-brick houses of the old city.

Kuwait is not, of course, alone in its contrasts. It would be a dull country that could not boast variety of scene and mood. But Kuwait's go deeper than most. It has always looked two ways – across the desert towards Mecca and out into the Gulf. Its present is highlighted by the confrontation of past and future; of a past which made a singular virtue of faith and a future which is pointed unhesitatingly towards technocracy and material well-being. Its merits and its frailties (for like every nation it has both), have their origins in the early history of Arabia, even in its prehistory; in the Islamic faith and the character of a desert race; in the political stances of imperial powers; in the fuel hunger of the world and the ingenuity of present-day technology.

One of the State's first planning advisers, the late Dr Saba Shiber, put it thus: "Yesterday there was old Kuwait; today new Kuwait. Yesterday and today are literally a stone's throw away in space, and fifteen years apart in time. The story of pre-oil and post-oil Kuwait is extraordinary and unique."

Geographically, the country consists of a strip of land measuring 7,400 square miles in area (including the Kuwait half of the so-called Neutral Zone), to the north-west of the Gulf (which Persians and others call "Persian" and the Arabs "Arabian"), with its township, its port, industrial and administrative centres on the south side of a naturally protected bay.

In broad outline, the political and social facts need little qualification. Kuwait's political foundation dates from 1710 when the Al Sabah family, descendants of the beduin Atib or Utub clan, who were of the Dahamshah branch of the Amarat tribe, left their homeland in the inner Najd and settled in this relatively hospitable coastal strip. The first recorded act of government was in 1756 when Shaikh Sabah I negotiated with the Turkish Governor at Basra for the independence of the territory. Thus, the rule of the Sabah household in Kuwait is sometimes said to proceed from that date. Nonetheless, bedu tribesmen had inhabited the region for centuries before its ruling family came from the Arabian interior to give it voice and cohesion.

After the rule of the Sabahs was established, Kuwait developed with a marked instinct for self-preservation, first under the suzerainty of the Turks and, from 1899, under British protection. In 1961, a Treaty of Independence abrogated that alliance and gave Kuwait freedom to play a sovereign role in the affairs of the Arabian peninsula and the world at large. In 1962, a draft constitution was ratified by the ruler and the country's first general election followed in January 1963. A census carried out in 1970 revealed that Kuwait's population had grown to 733,000 (it was estimated to be 206,000 in 1957 and 100,000 in 1939), of whom 346,000 were nationals and 387,298 expatriates of one kind or another. The country's gross national product was an estimated KD793 (£933 million) in 1968, or about £1,330 for every man, woman and child domiciled in the state.

But such facts present a superficial picture. It is in the dimmest recesses of the past, and in the long subsequent history of an Arab people of the desert who migrated to find here a good fresh water supply and the finest natural harbour of the Gulf, that the story of Kuwait resides.

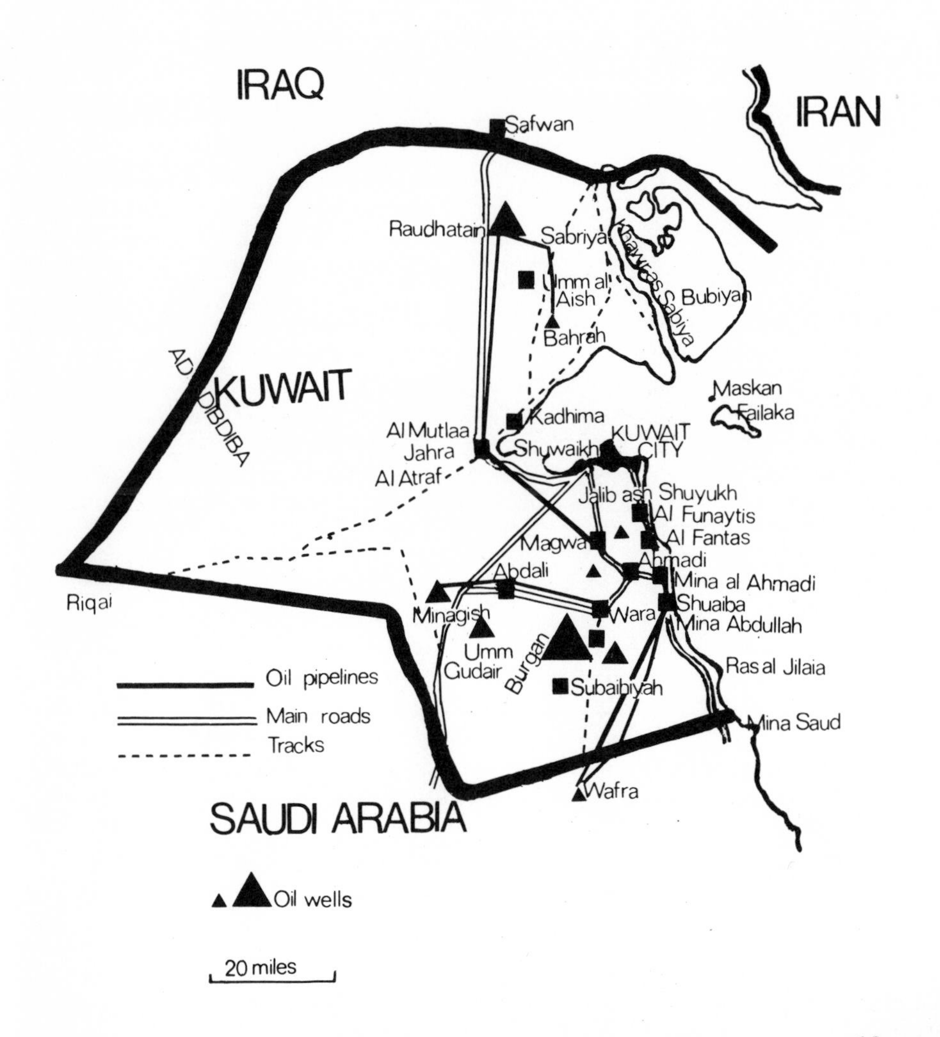
IRAQ
IRAN
Safwan
Raudhatain
Sabriya
Umm al Aish
Bahrah
Bubiyan
KUWAIT
AD DIBDIBA
Kadhima
Al Mutlaa
Jahra
Al Atraf
Shuwaikh
KUWAIT CITY
Maskan
Failaka
Jalib ash Shuyukh
Al Funaytis
Al Fantas
Magwa
Ahmadi
Abdali
Mina al Ahmadi
Shuaiba
Mina Abdullah
Riqai
Minagish
Wara
Umm Gudair
Burgan
Ras al Jilaia
Subaibiyah
Mina Saud
Wafra
Oil pipelines
Main roads
Tracks
SAUDI ARABIA
Oil wells
20 miles

Contents

Illustrations

MAPS

1. The Sands of Time

By geological standards, those events which had the profoundest influence on the territory of Kuwait took place in comparatively recent times – between 130 million and 70 million years ago. The mesozoic era, during which most of the world's great oil deposits were formed, stretches back 225 million years to the time when the earliest known reptiles appeared. It came to an end about 70 million years ago with the extinction of unimaginable hordes of the most monstrous creatures ever to inhabit the earth. That era embraced the triassic, jurassic and cretaceous periods, and it was during the last of these, in a time span of about 60 million years, that the foundation of Kuwait's vast oil deposit was laid. Then, according to the fossil evidence preserved in rocks, large areas were covered by vegetation and dominated by giant reptiles such as the ichthyosaurus, immense lizards and serpent-like creatures. Forests, swamps and seas teemed with animal and vegetable life. The seas flowed over the land, across the plains of Arabia, lapping up to the hills of the north, the west and the south. Then, after further long intervals, the seas ebbed. In this alternating advance and withdrawal of the sea vast quantities of plant life were both generated and destroyed. It is known that an ice age occurred in the cretaceous interval of time. A vast blanket of ice advanced across much of the world, from Europe to farthest Asia, and then gradually retreated. It may have been such a layer of ice which finally engulfed the last of the giant reptiles, for by the end of the cretaceous period they were extinct. Embedded in chalk and rock, they decomposed along with a squelching mass of plants and smaller creatures, the organic basis of the oil hydrocarbons to which intense heat and immense pressures reduced them over ensuing millions of years.

A sketchy explanation. But necessarily so, for the processes of natural change occurring over hundreds of millions of years do not resolve themselves simply or lend themselves to easy scrutiny. Much remains conjectural. The most that the geologist or the palaeontologist will commit himself to is the proposition that a transformation of the complex organic chemistry of life into simpler hydrocarbon forms, over immense periods of time and at great temperatures and pressures, is as plausible an explanation as can be advanced for the existence of oil. At any rate, it is an academic question. The liquid bounty is there, and for the Arab that is a matter more for gratitude than for argument as to why and wherefrom.

The evidence of depth and of rock formation suggests that much of Arabia's oil was, in fact, formed at different periods of the earth's formative upheaval. The plains of the great peninsula – framed by the land masses of Lebanon and Syria in the north, and by the highlands of the south and west – form an immense structural basin in which changing patterns of climate and life were able to interact and ultimately to produce some of the richest of all the world's oil deposits. Before the cretaceous period, even before the mesozoic era, in the so-called paleozoic segment of time which ranged over something like 350 million years, the same coming and going of seas and ice blankets, the alternating intervals of life generation and life extinction, took place. And, thus, by the transformation of unimaginable quantities of plant and animal life, oil was produced at other levels. These processes also went on after the main Kuwait deposits had been formed, into the eocene and miocene periods of the more recent tertiary or cenozoic era. From Iran and Iraq, down to the Gulf regions of Kuwait, Bahrain, and Qatar, beyond to Oman and Dhufar, landwards to eastern Saudi Arabia, oil-bearing sands and rocks testify to a passage of time which all but defies human comprehension, and to events whose major significance has been concealed from mankind, as though by an ingenious conspiracy of nature, until its moment of need.

By comparison with the geologist's time scale, the history of man is short measure. Nevertheless, the earliest inhabitants of Arabia probably occupied a land far removed in structure and climate from its present form – green and deciduous, well-watered and teeming with animal and plant life. Again, supposition must take the place of evidence. Curiously, less is known of man's earliest years on earth, than of the millions of years which preceded him. He left nothing so useful by way of evidence as the fossils, the encrusted plants, bones and molluscs, which enable the palaeontologist to make at least an inspired guess at time and circumstance. Arabia was probably on the edge of that last great ice sheet which covered much of Europe, Asia and America for nearly three million years, and which finally petered out some 11,000 years ago. As the ice advanced and retreated, perhaps four times during that expanse of time, grassland changed gradually to desert, running streams to dry wadis. The transition was slow, but in the end it gave rise to conditions of climate and terrain which are at the root of the Arab character, at once catalyst and inhibiting force in a remarkable historical development. To say that the Kuwaiti is an Arab and that his country's history is an integral part of Arabia's, coloured perhaps by a long association with the sea and maritime tradition, is to state the essential facts of the matter. But it begs a great many questions. We know, or at any rate can deduce, much that occurred in the vast desert expanse of the Najd and Hejaz over the 1,300 years or so since the dawn of Islam. But what of the millennia of human occupation before? What of the thousands of years between the first communities of the peninsula and the enlightenment; between the common ancestor of these peoples, Shem, son of Noah, and the Prophet Muhammad? And what of the even more dimly lit regions of the stone and iron ages? Frankly, we know little, though chinks of light have penetrated the darkness in recent times. For the rest we must fill the gaps in our knowledge as best we can by reference to the earliest communities and civilisations along the edge of the desert. Though the nomadic Arabs of the interior gave much to the places in which they settled and to the people they mixed with, they have left no tangible evidence of their own distant past.

The tens of thousands of years which separated the primitive tribes of the Arabian peninsula from the men who first learned the rudiments of the languages we came to call Semitic, are recorded by nothing more substantial than a question mark. It can reasonably be assumed that after the formation of the last great mountain ranges in Europe and Asia Minor, and after the glaciers had melted away, the earliest inhabitants of central Arabia roamed a wet and marshy territory, armed with the crudest implements of wood or stone. As the area became drier and almost bereft of plant life, these people must have gradually turned their thoughts and implements to husbandry of the soil and the herding of animals, learned to seek out the land most able to support crops and to cultivate it; to breed and shepherd flocks; to fashion more efficient instruments for work, and doubtless for war. Like other formative races of mankind, they began to produce basic linguistic forms. Some settled where there was water and sustenance. Others kept up a nomadic existence, taking their herds and their flocks with them. While most if not all other inhabited parts of the world remained in a state of savagery, settled communities formed in the basins of the Tigris and the Euphrates and it was there, especially towards the convergence of those rivers in the southern lowlands of Mesopotamia, that civilisation first began to take shape. Abraham was born 2,000 years before Christ, yet his birthplace Ur had seen near miracles of construction and civilised achievement during a thousand years before that. It was the Sumerians beyond the northern extremity of the Gulf who invented writing and gave the first real impetus to the advancement of mankind. Inland at Ur, Erech, Lagash and Nippur, and at Eridu near the coast, they developed the arts of pottery and sculpture, of architecture and vehicle design; they plotted the heavens and devised the first principles of astrology; they developed the sexagesimal method of numbering which survives today in the division of the hour into 60 minutes and the minute into 60 seconds, and in the geometric division of the circle into 360 degrees; they learned to irrigate the land; and while all this was going on the cities of Sumer fought pitched battles with each other. They exemplified at the very moment of man's emergence from primitive into civilised social organisation, the perversity that has dogged his progress ever since, the almost interdependent abilities to create and destroy, the limitless contradiction of a capacity to love and to hate in about equal measure. The Sumerians recorded their findings, their conflicts, their dead – and the names of the royal servants they buried alive so that they could continue to serve their kings in the life after – in cuneiform script on tablets of clay. Thus, archaeologists have been able to piece together a well documented account of this fertile strip of land from which some of the most majestic achievements of civilisation derive. To the north of Sumer, Semitic groups who long before had come as nomadic tribesmen from the Arabian desert, formed the kingdoms of Kish and Mari. Under the leadership of Sargon, they spread their dominion over an area from Elam, east of Sumer, to the Mediterranean Sea.

The Empire of Akkad embraced the whole of the region that was to become Babylonia, and much of Assyria. It was, in fact, the first empire of history and it almost certainly stretched as far as the area of the Gulf to the south of the valley of the Tigris and Euphrates. It was a relatively benevolent imperial power, for when the authority of the Akkadians declined it gave way to the combined rule of Sumer and Akkad, to a Sumerian culture regenerated by Semitic influence. By 1900 BC, another Semitic people – the Amorites – had taken over Babylonia and ousted from power the Akkadians and Sumerians, who nonetheless remained a powerful force in the area. Under the Amorite king Hammurabi, the regions of Assyria, Akkad, Babylonia and Chaldea, from beyond Nineveh in the north to the coastal tip of the Gulf in the south, became thriving centres of learning. They systematised the legal codes built up over several centuries by the Semites and Sumerians, and developed the arts and sciences through a virtually uninterrupted period of two centuries.

Then came the Hittites. Descending the Euphrates from Asia Minor, they occupied Assyria and Babylonia, thus resolving a conflict which raged almost without pause between the kings of those nations. Eventually the Hittites' influence spread across northern Arabia to the Mediterranean. They captured northern Syria and concluded a series of non-aggression treaties with the Egyptians, though

ii Ancient civilisations and places

neither party was thereby inhibited from attacking the other when opportunity offered. There was a constant flux of invasion and counter-invasion across the fertile plains of Assyria and Babylonia. By 1750 BC the Kassites from the region of Elam had established the third dynasty of Babylon, their kings ruling for 576 years. In 1175BC Nebuchadnezzar came to power to face continuing conflict with Assyria and a new onslaught from the Arameans. For another 500 years, with only brief periods of respite, Babylon was weakened by invasion and internal dissension, prey to the Assyrians, Arameans and Elamites. Meanwhile, Assyria became revitalised and, combining a culture derived from Babylon with a natural instinct for commerce and combat, spread learning, trade and disarray throughout the Middle East. They deported entire conquered tribes by way of upholding their rule. Yet they also created some of the finest buildings of the ancient world and, at their capital of Nineveh, built up the most famous library of antiquity. When Assurbanipal, the last great king of Assyria, had completed his campaigns agains Egypt, Elam and Babylon between 668 and 626 BC, the energies of his empire were exhausted. It was eventually brought to ruin and divided by the re-emergent Babylonians and Medes.

The Chaldeans, people of Arabic and Aramean stock, had begun to appear in Southern Babylon in about 1000 BC. They were a constant thorn in the side of Assyrian conquerers and it was under their influence, and under the leadership of Nebuchadnezzar II, that the neo-Babylonian Empire was established, bringing a new if brief splendour to the region marked by the confluence of the Tigris and Euphrates. The sacred date tree flourished in fertile, well-irrigated plains, the hanging gardens and the great walls were built, astronomy and the mathematical sciences were given a new lease of life. But Chaldean Babylon learned no more of the art of conserving its energies by peaceful deployment than did its predecessors or successors. It destroyed Jerusalem and its army even crossed over to Egypt.

When Nebuchadnezzar died, Babylonia began its final decline. Under Nabonidus, it fell to the Persian army under Cyrus without protest, its intellectual and material splendours gradually fading under the successive rule of Persian, Greek, Parthian and Sassanid, until the Arab caliphate established in AD 636. Such were the fortunes and achievements of the people of the arc which sweeps upward and across from Gulf to Egypt. While they held sway, creativity and scholarship thrived.

The cross-currents of dominion and language and custom in these fertile lands gave rise to immense intellectual and material progress. We know little of what went before. When archaeologists began to uncover the evidence of habitations of 4,000 years or more ago, there were few links with the prehistoric past. Unlike Egypt, where flint implements and metallic objects traced an evolutionary path from primitive bush life, through a neolithic period to the first dynasties, the ancient sites of Sumerian and Semitic occupation revealed few such traces. It must be assumed that most examples of palaeolithic or neolithic workmanship were removed by flood from the delta region of the Tigris and Euphrates, before the first cities were built and the land reclaimed. Recent archaeological discoveries of the greatest importance in the region of the Gulf make up for much of that deficiency in our knowledge, but more of them later.

If the progression from prehistoric to historic man has for long been obscure, more recent developments along the periphery of Arabia are well and truly recorded, or at any rate implied, in the documents of the three great religions which have their origins in the Semitic lands. Community life and trade prospered for three thousand years before Christ in the crescent area which began just below the place that is now Basra at the tip of the Gulf, traced an arc through Babylonia, Assyria, Phoenicia, Syria and Palestine, and terminated at Aqaba in the west. Settled communities also formed along the highlands that skirt the Red Sea, from Jiddah to South Arabia, and along the Arabian Sea coastline to Dhufar. Trade routes brought the inhabitants of these distant places into contact with each other and with people of the desert.

The overspill from the deserts regularly invaded the settled areas whose less warlike people fell easy victims to the nomads. The influx of such tribesmen into the settled areas infused them with new blood and new virility. Sometimes they simply bartered their wares of livestock or fur, skins, silk, wool, spices, ivory, frankincense, myrrh or

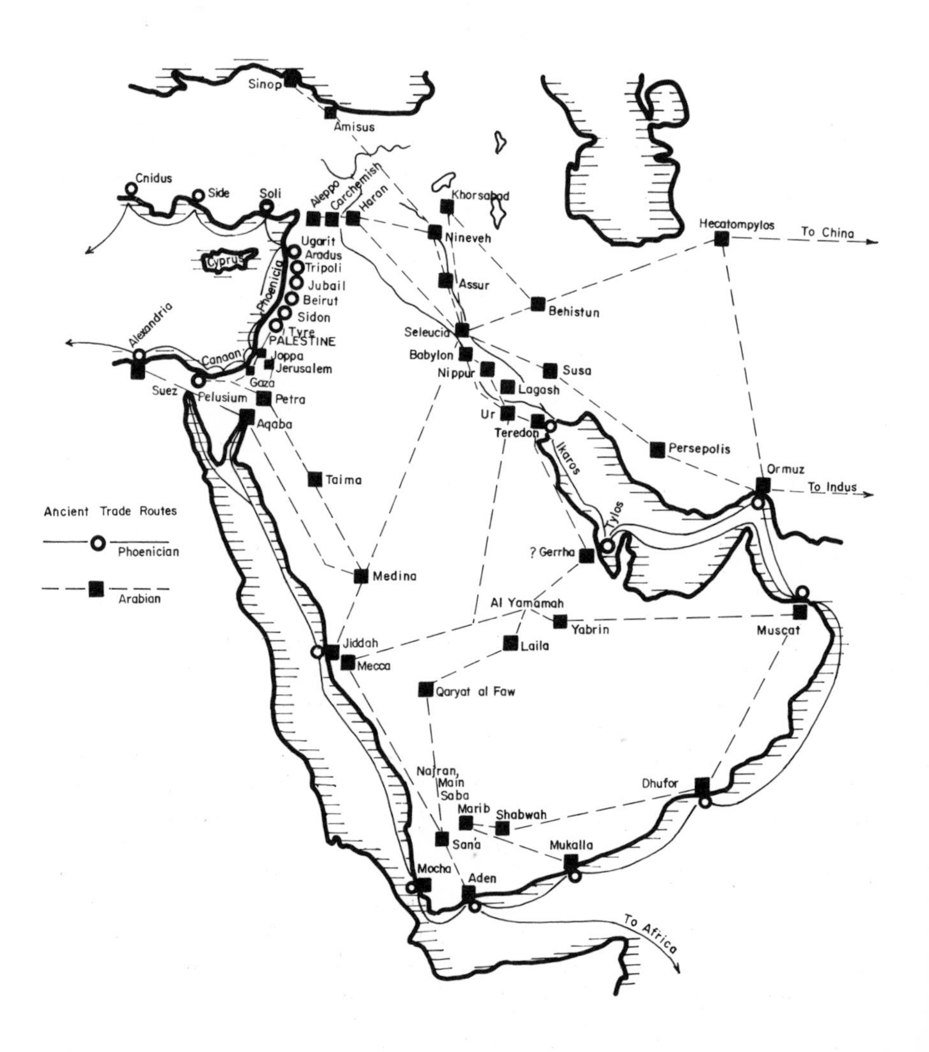

iii Ancient Trade Routes

fuel, and returned from whence they came. From Marib, Mukalla, Dhufar and other southern places, from Muscat, Mecca, Hufuf and Al Yamamah, the caravans of trade made their way along the coast or across the desert. Baghdad, Hail, Basra and Medina were the staging posts on the roads to Samarkand, Carchemish, Damascus, Joppa, Jerusalem, Alexandria and Petra. Their goods found markets in Cappadocia and the Hittite cities of Sinop and Amisus on the Black Sea; in Persia and along the Caspian seaboard, even in China. The people of Arabia went to Suez and into Egypt, across land and sea to Ethiopia, India and Byzantium. Their influence on the people around them was widespread and lasting. When they embraced the faith of Islam, that influence took on a new dimension. Impelled by a zealous trading instinct and given impetus by the words of the Prophet, they conquered a large part of the world, in spirit and in deed. But though the Arab has always been extrovert in his trading habits and in his determination to spread the teaching of the Prophet, he is in the very basic sense of national character, inward-looking. The people of the Arabian peninsula have retained a remarkable identity of speech and literature, and nowhere has the purity of their language been so well preserved as in the desert. In the thousands of years during which coastal Arabs have been in communication with the world at large, corrupting their speech and the written word as all languages are corrupted by trade and the domination of foreign powers, the Arab of the hinterland has kept his customs and his tongue remarkably intact; and when it has become necessary he has revitalised the habits of the townspeople on the periphery. If the term Semite is but the approximation of a German scholar in defining a race, it serves well enough. The descendants of the biblical Shem are, with all their differences, more alike in essential qualities than any other people of comparable dissemination, save perhaps the Chinese.

2. In Search of a Lost Past

We arrive at a point of diversion; a point at which much that has gone before must be qualified, or even contradicted.

The emergence of the first civilised communities in the region of Mesopotamia is no matter for conjecture. The cuneiform tablets of Babylon and Assyria, buried for 3,000 years and more, have been excavated and examined (though several thousand have yet to be translated), and made to divulge legends and historical facts from which a vast body of knowledge has been derived. In the absence of material evidence of older or even contemporary habitations, the great cities of antiquity which arose along the rivers Tigris and Euphrates, along with those of the Nile Valley, have become the accepted birthplaces of civilisation. But were they?

In 1953, a Danish archaeological expedition began an investigation in the area of the Gulf which, after fifteen years of patient digging and sifting, began to throw a startling new light on the discoveries and conclusions of the previous century, and to involve its founders in major questions of scholarship. Much of the evidence produced by the digs of this Danish team has been published by Geoffrey Bibby, an Englishman who joined the Danes as field director of the expedition, in his book *Looking for Dilmun*.

Looking for Dilmun meant, in fact, looking for a lost community—perhaps a mythical one. And like all such ventures it appealed more to those who wished to embark on it than to the institutions asked to provide the money. Lost civilisations represent a popular cause among enthusiasts and adventurers, but they seldom appeal to universities, oil companies or governments. Thus, the expedition led by Professor Peter Vilhelm Glob, head of the Danish National Museum, concealed its real, if somewhat nebulous, ambitions beneath a plausible cloak. They were, they said, going to look around the famous burial mounds of Bahrain, to search for clues that might lead to something in the graves that cover much of the island like overgrown anthills and which, over the centuries, have invited much curiosity.

They were armed with some intriguing facts and even more intriguing theories.

In 1925, at about the time that Sir Leonard Woolley and Sir Flinders Petrie were digging at Ur and Babylon and Nineveh, Ernest Mackay went to Bahrain to examine the burial mounds which earlier travellers had described and even excavated, though usually with more intuition than skill. Mackay described them in considerable detail. Each sand-covered mound was a man-made chamber of stone with provision for single or double burials; some were double-deckers. Many contained damaged or fragmentary skeletons, others just a few bones. Occasionally he found half ostrich eggs, presumably used as drinking vessels; some fragments of pottery were evident. It was, said Mackay, a necropolis, a land of the dead, in which were buried nomadic Arabs of the mainland, from about the second millennium before Christ to more recent times.

Other visitors held conflicting theories. In 1906, Colonel Prideaux of the Indian Army went to the village of Ali, where giant mounds thought to contain the royal dead rose up in the rear of thousands of lesser graves. He dug with great energy and resourcefulness, unearthing gold rings and two ivory statuettes. The latter were examined by British Museum experts who declared them to resemble existing ivories thought to be of Phoenician origin, thus lending weight to the remark of Herodotus that the Phoenicians claimed ancestry in the region of the Arabian Gulf.

But it was another amateur archaeologist, Captain Durand, who provided the Danish expedition of 1953 with the substance of its suspicions. In 1897, the British Foreign Office, ever astute whatever the mask of naivety it chooses to wear, sent Captain Durand to Bahrain to prepare a political report; archaeology was his cover. He had a keen eye. Sites outside the burial ground would, he suggested, repay investigation. More importantly, he returned home with a black basalt stone which bore an inscription in cuneiform, the wedge-shaped script which philologists were beginning at about that time to translate, more than two centuries after the first copies of the Persepolis inscriptions had been brought to Europe from the capital of Darius of Persia. It read:

> Palace of Rimum,
> Servant of the God Inzak,
> man of Agarum.

In 1880 Sir Henry Rawlinson, perhaps the most renowned of all the great Assyrian scholars, wrote a commentary on Captain Durand's survey of Bahrain antiquities. By then familiar with the epic poetry and history of Assyria from the tablets in the British Museum, he commented: "Throughout the Assyrian tablets, from the earliest period to the latest, there is constant allusion to an island called Niduk-Ki in Accadian, and Tilvun or Tilmun in Assyrian, and . . . . this name . . . . unquestionably applies to Bahrain."

That assertion made by a man who was not given to uninformed surmise, was very much in the minds of the Danish archaeologists when they went to Bahrain to commence their investigations. Nevertheless, they went about the business of looking into burial mounds, and maintained a discreet silence on the subject of the lost kingdom. The archaeologist's commonest pastime is the collection of pottery – or, in his own language, potsherds. Nothing, they say, is as effective a guide to the age and habits of a community as its potsherds. In Bahrain, they found plenty of fragments on the surface, but none of them were of significant age. The oldest were pieces from China of the Ming Dynasty, sixteenth–seventeenth century AD, and thus, in archaeological terms, modern.

The Danes decided to excavate a burial mound – distant from those investigated by Durand, Prideaux and Mackay, so that they could compare notes on the tombs and their artifacts. Like all the graves of Bahrain, theirs had been visited by robbers in the course of time, and nothing of use or value was visible when they cut through the mound to the burial chamber. But under some fallen blocking stones, used to shore up the entrance, they found the inevitable potsherds, and then a few bones. The pieces of pottery formed a tall, red-coloured vessel. The bones were arranged into the lower half of an adult human being, lying on its side, legs half bent. When they cleared a way to the end of a chamber, they found two copper spearheads, driven into cracks in the stones. The haul was interesting if only because it confirmed the archaeologists' suspicion that the contents of the grave mounds were totally unrelated to surface finds. They had accumulated potsherds by the crate; they had found flint instruments of Stone Age man, deposited perhaps 40,000 years before. That the pottery fragments and implements of the tombs were of a different age, there was no doubt. But surely the people who built those tombs, one hundred thousand or more of them, had lived somewhere on the island? Surely they had left some more evidence of who they were, and when and why they settled?

If there were no connecting links on the surface, perhaps they were hidden by the accretion of sand over the centuries.

They began again to dig. They sought out the sites which showed signs of having been artificially raised in the course of time, one settlement being built on top of another, because the site was usually the best that was available in the proximity of water. They dug in two places. One at Diraz in the north-west, a circular hollow surrounded by sand-covered mounds, which in turn were surmounted by large blocks of stone. The other, to the east of the first site, by the village of Barbar, a large mound. Professor Glob (universally known as PV) noticed two large blocks of limestone protruding from its northern slope, thus distinguishing it from the large burial mounds nearby. Further digging revealed that the blocks had sides measuring four feet; they weighed over three tons each.

It is no part of our purpose to go into the complex details of archaeological excavation. Indeed, Mr Bibby's own account of the expedition is so graphic in its detail and so comprehensive in its presentation of the wide-ranging issues of scholarship involved, that to attempt any more than a summary of the story here would be an impertinence. It is the discoveries that matter in the context of Kuwait and its relationship to the history of Arabia. And at Diraz and Barbar, the results of the digs conducted by the Danes were to set in motion processes of comparative study, of recollection and deduction which have changed our vision of the earliest history of mankind.

At Diraz, they found a flight of steps leading from the surrounding mound to a well, its head flush with the floor of a tiny, stone-built chamber. There, they found two figures of kneeling, headless quadrupeds, a piece of alabaster bowl, and thirty or more potsherds. All were labelled and packed away. They were not looked at again for two years. Nobody had noticed that the pottery, red in colour with horizontal ridges, was typical neither of the grave mound nor Islamic-period potsherds so far recovered.

For some time the Barbar mound proved unrewarding. The Danes dug to lower and lower levels, uncertain whether they were contending with a very large burial mound or with a "tell", an Arabic word used to describe the building of one habitation on the derelict remains of another, often resulting in cumulative settlements over hundreds or even thousands of years, on the same spot. Eventually, they looked down upon the court of a temple with features remarkably similar to those depicted by the famous seals of Mesopotamia – the seat of a god with an altar in front of him, with supplicants bearing offerings. The centre of the temple at Barbar consisted of an oblong framing two circles. In front of the altar was a pit which, when it was excavated, gave up scattered hauls of lapis-lazuli beads, vases, a copper figure of a bird and a statuette of a naked man in an attitude of supplication. A few enigmatic pieces – arms and other parts of statues – were found in a rubbish heap, the remnants it seemed of an earlier temple that had been thrown out when the later structure was built.

The little copper figure of a man was Sumerian in appearance. One of the alabaster vases, too, was of a shape recorded from the Mesopotamian finds. Had they been found in Babylon or Assyria, there would have been no hesitation is ascribing them to a period between 2500 and 1800 BC. This time, note was made of the fact that the pottery finds, thin red ware with low horizontal ridges, were untypical.

A temple with Sumerian characteristics had been found some 400 miles from the head of the Gulf, where Mesopotamian civilisation was thought to have ended. It was a find of some magnitude. But a great deal of reference back to the libraries of Assyria and Nippur, to seals and inscribed tablets of the known civilisations, was needed before these discoveries could be made to fit with any accuracy into the story which, according to all known records, began in Sumer and Akkad about 2800 BC and ended with the destruction of Assyria around 700 BC. There was also a lot of digging still to be done.

Nevertheless, the name Dilmun was by now in common use among the archaeologists and in their conversations with the far-sighted ruler of Bahrain, Shaikh Sir Sulman bin Hamad Al-Khalifah, historically the friend and ally of the ruling household of Kuwait.

It was in 1872 that the word Dilmun began to see the light of day after an obscurity of more than 2,500 years. Then, from the thousands of cuneiform tablets of Assyria's royal library, an account of the great flood, the universal deluge, was published. It was the eleventh chapter of the story of Gilgamesh, mythical king of Erech, and told of the ancient Utu-nipishtim, the survivor of the flood; of how when the gods decided to destroy mankind, Enki, god of the waters under the earth, told Utu-nipishtim to build an ark and take his family and possessions aboard it; of the old man's sacrifice to the gods and Enki's intercession that the gods might never again punish all mankind for the sins of some. Enlil, the great god, agreed. He went aboard the ark and proclaimed: "Hitherto has Utu-nipishtim been mortal. Now shall he be like a god, and dwell in the distance, at the mouth of the rivers."

The affinity with the biblical story of the great flood was obvious. The world began to take notice of the Assyrian tablets from Assurbanipal's library. Nearly thirty years later, in 1900, an American expedition from Pennsylvania University unearthed the Ziggurat of Nippur, a mound with a temple at its summit. At its foot lay the temple of Enlil, the great god of the epic of Gilgamesh. Within the archives of that temple 35,000 cuneiform tablets were found, many of them in Sumerian. Such a number of tablets could hardly be translated overnight. It was not until 1914, in fact, that a Nippur inscription was translated giving a very much earlier Sumerian version of the story of the flood. The tablet was not complete, but those parts that survived are of vital importance to our story. In the Sumerian version Utu-nipishtim was called Ziusudra. The last part of the text, in the translation of Professor Kramer of Philadelphia, reads:

"Anu and Enlil cherished Ziusudra, life like a god they give him, breath eternal like a god they bring down for him. Then Ziusudra the king, the preserver of the name of vegetation and of the seed of mankind, in the land of crossing, the land of Dilmun, the place where the sun rises, they caused to dwell."

The Sumerians described the Gulf of Arabia variously as the Lower Sea, the Bitter Sea and the Sea of the Rising Sun.

Though the actual location of Dilmun remained in question, there could be no doubt that the Sumerians, the oldest inhabitants of the plains of Mesopotamia, knew of another civilisation. Nor can there be doubt as to the fact that it was in the direction of the Gulf.

Other tablets from Nippur, translated many years later, brought to light the mythical poem of Enki, guardian of the fresh water, and Ninhursag, goddess of the land. It opens, again in Professor Kramer's translation:

The holy cities – present them to him,
The land of Dilmun is holy.
Holy Sumer – present it to him,
The land of Dilmun is holy.

The existence of Dilmun was clear, once the tablets of Nippur and Assyria, separated by nearly two millennia in time, had become part of recorded history. Meanwhile, by 1955, the Danish archaeological team had unearthed a temple whose features and contents bore a marked resemblance to some of the finds at Babylon and Ur.

And while these initial discoveries were being made, the Danes were also digging elsewhere. They chose the Portuguese fort, as visitors and passing sailors had called it for centuries past, though they called it Qala'at al-Bahrain, the fort of Bahrain. And after another long period of patient, painstaking digging in the sweltering heat of the Gulf, they made another find of immense importance. A city of nearly forty acres, lying below the fort, within shovelling distance of its moat, two-thirds the size of Ur of the Chaldees which Sir Leonard Woolley had dug for twelve years without exhausting its possibilities. As they cut through the earth they came to two massive stone walls, abutting at the northern edge of their hole. Eventually, they came to a cement floor, pierced by two large oval holes which further digging showed to contain thick, earthenware bath tubs, coated on the inside with bitumen – or that, at any rate, was what they looked like. When they cleared away the debris, however, a huddled human skeleton was found in each. They were, of course, sarcophagi, and their skeletal contents were clearly not Muslims since they were not oriented towards Mecca. Beneath the floor there were two other base levels. They dug even deeper in the small space between the ancient stone walls and a southern mound, and their labour was rewarded by a few potsherds of "respectable age". Flakes of flint suggested neolithic habitation, beyond 3000 BC. Questions were piling up again. Nothing could be dated with certainty.

By good fortune, the sea came to the rescue of the archaeologists. Waves advanced towards their city and uncovered a stretch of masonry, exposing the boundary wall. They began to delineate streets and habitations, crazy paving and well-engineered drains; the town was completely symmetrical, even to the sitings of the doorways of the houses. It was interesting, but in archaeological terms something of an anticlimax. The Danes and their British field director had, it transpired, been digging an Islamic fort, most of the contents of which were disappointingly recent. Potsherds dating back to about the fourth century BC were the only noteworthy advance on Chinese porcelain of perhaps AD 900 and Attic ware that might have reached back to 200 BC. They had hoped for Barbar sherds, thin, red sherds such as they had found in the still progressing Barbar dig. Those pieces of pottery, together with a bull's head and the diminutive Sumerian man took them back to 2300 BC or even before. At the fort they remained, impatiently, within the last two millennia. They decided to dig a new hole, twenty-five metres away, while they contemplated the nature of their haul so far.

For the sites at Barbar and Ali and at the fort, they had arrived at dates varying from 2300 BC to AD 900, largely on the evidence of pottery finds, showing beyond doubt that a very ancient culture had thrived hereabout and that subsequent occupants of the area had traded as far afield as China. But the picture was by no means complete or entirely coherent.

They dug down through seven city sequences, each revealing its own particular characteristics, some containing the remnants of earlier or later habitations; all had been robbed and ransacked at some stage. The pieces left behind for the Danish expedition to construct their jig-saw from were barely adequate. But they made the best of what there was.

At the fort, they found a third bath tub. Surprisingly this latest sarcophagus had escaped the attention of robbers. It contained a skeleton as before and a complete bronze drinking service. There was also a wine jar and a glazed pottery vessel. They decided that the skeleton was of a man, buried with an iron dagger at his waist and a seal of agate around his neck. However, this burial chamber turned out to be only an alcove below the stairs of a large hall. The architectural detail of the hall was considerable, yet it could not be dated by reference to any building of which the expedition had knowledge. Pottery of the age of Alexander was found. Had the coffins been placed in position during the occupation of the palace? The practice was not unheard of in Mesopotamia. Then, another piece of detection led them a stage further. Several light patches were noticed in the floor. Probing revealed twelve bowls from a total of fourteen holes. One of the holes without a pot was covered by a layer of broken pottery, below which was a collection of twenty-six beads of agate, amethyst and faience. The pottery was considered to be contemporary with the palace, probably the Assyrian period of Sargon, somewhere about 700 BC. But there was still a lobby among the archaeologists for a date as far back as the Kassite ascendancy, beyond 1200 BC, while some inclined to King Uperi of Assyria in 709 BC. Whichever prediction was borne out, it merely demonstrated that there was more

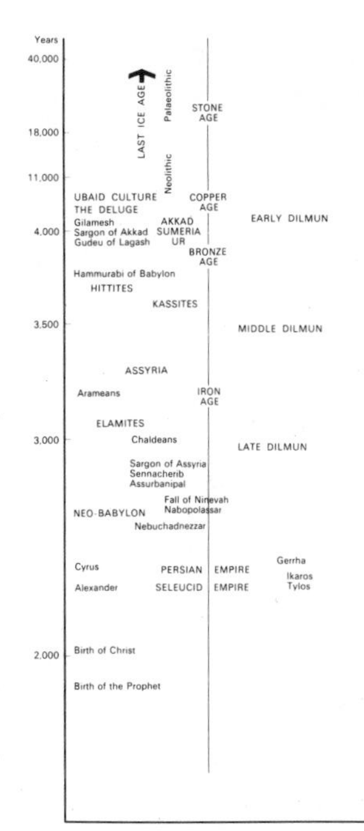

2. Archaeological time scale

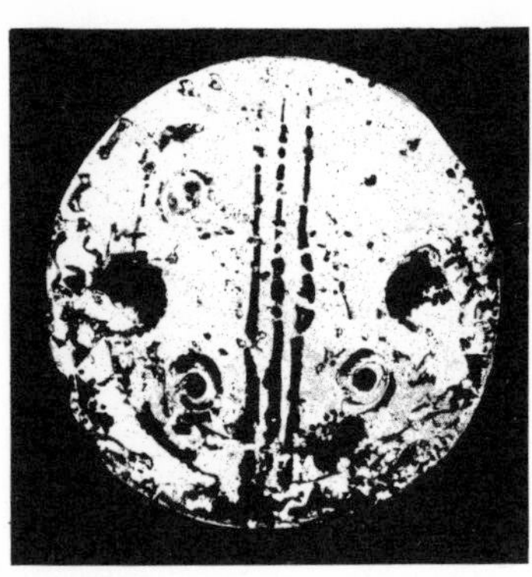

3. Reverse side of a round Dilmun seal, with three incised lines and four circles

4. Dilmun's seals are characteristically round, while those of the Indus civilisations are generally rectangular, see illustration opposite. But the bull is a recurring feature of both. Some Dilmun-type seals were found at Mohenjo-daro and vice-versa

searching, much more, still to be done. They were sure by now that they were in Dilmun, though they were far from knowing its confines of time and space. They did recall, however, that the first recorded mention of Dilmun was in one of the tablets of Sir Leonard Woolley's dig at Ur, the tablet of Ur-nanshe, King of Lagash, who lived about 2500 BC.

"The ships of Dilmun . . . brought me wood as tribute."

They looked more closely at the contents of their latest pottery vessels. They found skeletons and loose bones of snakes and, in most cases, a tiny turquoise bead. It tied up with the legend of Gilgamesh; it was perhaps here that one of the stories told in that legend – of the snake that ate the pearl and achieved immortality – began. Except that a turquoise is not the same thing, in value or prestige, as a pearl. A poor man's tribute perhaps?

The next find was of critical importance. It was a round seal of steatite or soapstone, an inch in diameter. Looking back, from the earliest levels of inhabitation to the latest, they were now at the stages of cities I and II on the fort site, and this first seal was found in a roadway at the level of the second city.

Thus, while one team was digging in, say, city V among the pottery and other objects contemporary with Alexander, another was working at a level anything up to 2,000 years older in its features and signs of habitation. It was not long before the men working at the most ancient levels were pursuing more seals in the direction of city I. They came to a level where the pottery bore a "chain-ridge" decoration. The city was unwalled and had been burned. From the evidence of consistent "chain-ridge" pottery, they were able to establish that it had been rebuilt by the same inhabitants. They found a second seal, and later, a third came to light. All were round, but otherwise they were perplexing in their differences. The first seal had a face design made up of two human figures, with a pierced dome on the other side, decorated with three incised lines and four circles, each with a dot in the middle. The second had a high pierced boss with a single incised line and on its face the figure of a bull. It bore an inscribed human footprint and a scorpion. The third was smaller than the others and of black steatite instead of grey. On the face was a goat or gazelle; above it another similar animal and a star. The archaeologists reflected again on the excavation at Ur, and on matters of identification which transferred their thoughts across the Gulf to the ancient cities of Harappa and Mohenjo-daro, on the banks of the Indus river. The civilisation of the Indus probably developed at about the same time as those of Mesopotamia. It too, produced seals and objects of copper and flint. It planned its towns with care and ingenuity. And its stamp seals usually contained a bull on their faces. The language of the early settlers of the Indus Valley is unknown, for they left no history or mythology, as did the Babylonians and Assyrians, on imperishable slabs. The seals of the Indus Valley find were mostly square, unlike the round ones of Bahrain, while those found by Woolley at Ur were generally cylindrical. It was, however, the exception that proved of uncommon interest to the Dilmun seekers. Three seals found at Mohenjo-daro were round. And so were a few of those found at Ur. Of the latter, seven known to have come from Woolley's excavations at Ur and one from another source, though possibly emanating from the same area, were almost identical to one of the Bahrain seals.

Thus, seals from three distinct and far removed regions bore an uncanny resemblance to each other. If one could be dated so could the others. Two of Sir Leonard Woolley's seals were of the Mohenjo-daro type, and were thought to be of Akkadian date, about 2300 BC, or perhaps the second dynasty of Ur, 2200 BC. Another was labelled Kassite, 1700–1200 BC. The civilisation of the Indus, whose language it has been impossible to learn because of the meagreness of the inscriptions that have survived, could now be dated to the last few centuries of the third millennium BC – to 2300 or beyond. The implications were of enormous importance in the attempt to an understanding of the Gulf civilisation which was emerging with increasing clarity from the work of the Danes. The Indus and Mesopotamian seals, remember, were foreign; of different shape and inscription from the great majority. They showed that there had been communication between these ancient people of India and Mesopotamia. But it was just as obvious that the seals had originated elsewhere.

At the point where they found the third seal at Qala'at al-Bahrain, they also discovered small fragments of steatite bowls, with the same decoration as the last two seals, and turquoise beads such as were in the pots of the burial chamber. There were also many small scraps of copper. Then they reached bedrock. But the detective work could go on by the incredible chance of Woolley's activity in 1930. At that time the great archaeologist of Ur decided to leave his work among the temples and palaces of kings and to have a look at the outskirts of the city, the residential quarter. He took with him a hundred and fifty workmen, and he uncovered among other things the dwelling house of Ea-nasir. Tablets found there showed that he was a merchant of moderate means, dealing mainly in copper. There were accounts of customers and quantities of goods, even letters of complaint. One of these remarked:

"When you came you said, 'I will give good ingots to Gimil-Sin.' That is what you said, but you have not done so . . . Who am I that you treat me so contemptuously? Are we not both gentlemen? . . . Who among the Dilmun traders has acted against me in this way?"

Many other references to Dilmun and the cargoes of copper, silver, pearls, precious stones, ivory, tortoise-shell and other commodities, point unerringly to Dilmun as the source of trade, and to Bahrain as, at least, a clearing house of that trade; for the ships that brought these goods from India and Afghanistan certainly came from that part of the Gulf, and Bahrain would be an ideal port. But what did the ships take away with them? Again, the evidence is provided by the tablets with which Mesopotamian civilisation was uniquely able to preserve the facts of its history. A receipt, issued by a certain Ur-gur and dated 2027 BC, records: "ten talents of different kinds of wool of ordinary quality, put in a boat to Dilmun." Wool, skins, piece goods appeared to constitute the return cargoes. One tablet referred to copper from Makan – another place of unknown location.

5. Indus square seal with bull impression

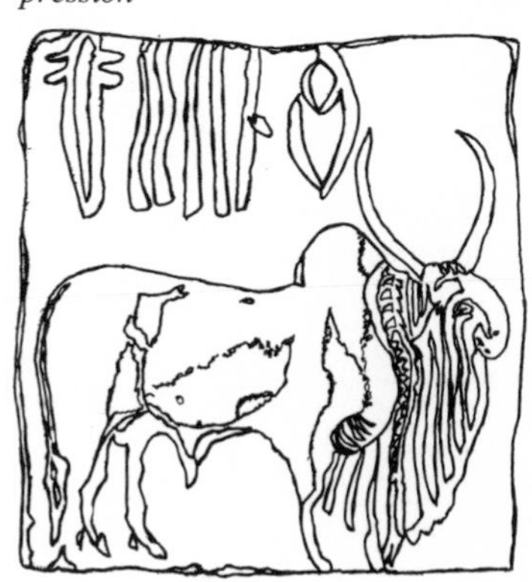

The chain of events had become impressive in length, if inconclusive in detail. The few scattered contents of burial mounds and palaces, objects of ivory and copper, the bills and receipts of Ur, the legends of Babylonia and Assyria, all pointed to a place of size and stature called Dilmun, in the direction of the Gulf, and to Bahrain as at least part

6. Part of a cylindrical seal commonly found in Mesopotamia

if not the entirety, of that place.

The burial mounds themselves were evidence of a community of no small size. The mounds at Ali, especially, could only have been built by a large work force over a long period. The city unearthed by the Danes was over half the size of Ur, and that was one of the largest cities of Babylonia. The seals of Bahrain were similar to those left behind by foreign merchants in Ur and the Indus Valley.

The evidence was more than circumstantial that a civilisation at least contemporary with those of Mesopotamia and India had existed in the Gulf. For the 2,000 years that separated Sargon of Akkad from Sargon of Assyria, perhaps beyond both into the third and first millennia BC, trade and creative activity had flourished by the Sea of the Rising Sun.

The Danish archaeologists were in no doubt that this at least had been established when, at the instigation of Britain's then Political Agent in Kuwait, Sir Gawain Bell, they went off to a part of the Gulf whose position brings us back to the mainstream of our story.

Dilmun's northern extremity

Failaka stands at the entrance to Kuwait bay, a small island which overlooks the mainland. A sentinel island. It was described by H. R. P. Dickson, during his term as Britain's Political Agent:

7. Smaller type of seal found at Dilmun level in Failaka

"Eight miles in length, uninhabited except for its western end, and a small village to the north having a dozen families. Extensive ruins", he said, "suggested a much bigger population at one time. Zor, the capital, was a flourishing little spot with good water. Clean, its inhabitants kind to strangers, a favourite place for the Shaikhs of Kuwait and their families when in need of a holiday. The islanders excelled at fishing and cultivation and they grew wheat among other crops." He added: "A feature of the island is the shrine of Al Khidr. This is a small shrine at the end of a small promontory. Although the orthodox Sunni Muslim will have nothing to do with the superstition, it is generally accepted by the islanders and by those of the Shia persuasion on the mainland of Arabia and Iran that Al Khidr pays a visit every Wednesday to the shrine *en route* to Mecca from his headquarters in Basra, and again on Saturdays when he returns. Hence any woman who is unable to have a child can, if she finds favour with the spirit saint, and is considered worthy, satisfy her desire. For she then miraculously becomes pregnant and all is well." Dickson concluded that though others associated Al Khidr with saints such as St George and the Prophet Elias, he favoured the Ishtar theory. "After all, Sumer and Akkad and later Babylon were close at hand, and Ishtar had an important part in their worship".

Bit-Iakin, conquered by the later Sargon, was on the shore of the Sea of the Rising Sun, "as far as the border of Dilmun" according to one of the Assyrian tablets. Where, then, were the borders of Dilmun? What was the extent of that lost land to which both mythology and factual documents of trade alluded more than 2,500 years ago? Where, indeed, was Bit-Iakin, land of the House of Iakin, which had earlier been called the sealand and had at times dominated Southern Babylon?

Failaka threw an almost immediate light on the matter.

Whereas the Danish archaeological team had been forced to adopt a rule of thumb in its Bahrain dig, carefully building a chronology based on well-concealed fragments of evidence – potsherds and skeletons, statuettes and pieces of metal – Failaka's clues were spread around its surface. Armed with the knowledge they had so painstakingly acquired at Bahrain, they had to do little more than bend down and pick up pieces of pottery in order to proclaim "Barbar" and "Greek". They identified, as much to their own surprise as anyone else's, two civilisations separated by some 2,000 years, the first of them going back to Abraham, the latter to Alexander the Great. More importantly, they had found by the testimony of the potsherd, habitations of the same kind and time as those of the fort and the Barbar temple of Bahrain. The breadth of Dilmun was becoming apparent. Yet Kuwait's diminutive island is closer to Ur and the other southern cities of Babylonia than to Bahrain. The distance from Bahrain to Kuwait, about 250 miles, is as great as that between the northern and southern boundaries of Babylonia at the height of its influence. Of course, first finds like these cannot be regarded as conclusive evidence of settlements. There was, inevitably, a lot of digging to be done. The year was 1957.

Pottery scattered in large quantities over the surface of Failaka was not the only favourable circumstance to greet the excavators. A brief look around showed a pair of well-defined "tells" practically next to each other, two accretions of buildings which, on this small but compact island could provide a direct route to two distant segments of the past. In Bahrain, though they had been digging for several years, they were confined to a few major sites in a relatively large area where sand probably concealed far more than had yet been uncovered. Failaka island, like so many things in Kuwait, was generous in its divulging of riches. And the government of Kuwait, keen to give practical and financial backing to the reconstruction of the country's past as well as to the construction of its future, made ample resources available.

8. Single seal with unread Indus script found at Failaka

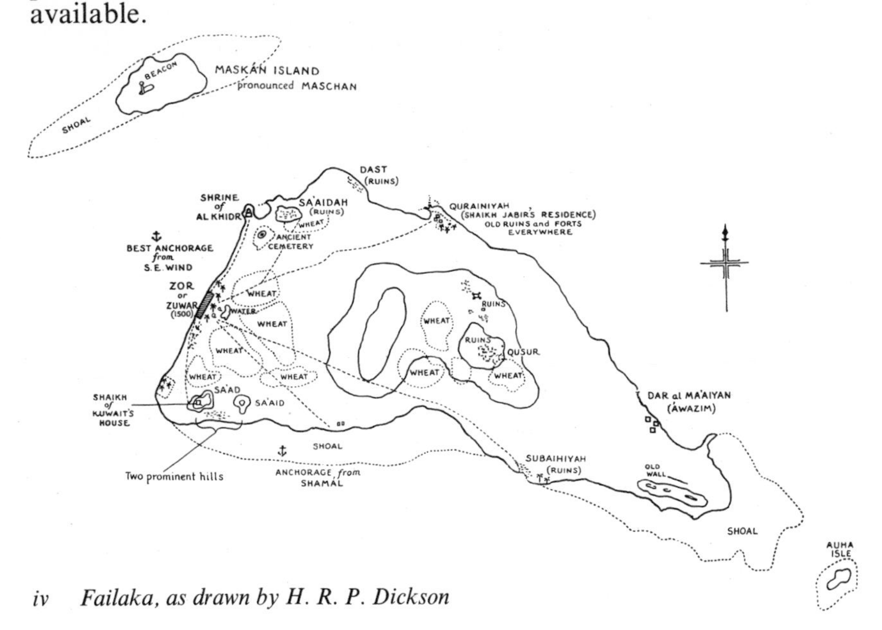

iv Failaka, as drawn by H. R. P. Dickson

During their first reconnaissance, the archaeologists noticed the shrine of Al-Khidr, the Green Man. It had been destroyed several times, but the inhabitants had always rebuilt it and their Shia brethren from far and wide, even as far as India, continued to visit it to pay homage to the Green Man, and to plead for the propagation of male children. Was its significance any deeper than the idolatry of which the Sunni Muslims of the mainland accused its adherents?

There were other matters to be investigated; other signs of antiquity which the archaeologists could, in the patient, time-spanning manner of their trade, turn over, weigh, compare and docket. Carsten Niebuhr, the great eighteenth-century explorer and cartographer, had declared that there was a Portuguese fort at Kuwait. It must have been on Failaka since there is no such fort on the mainland. Indeed, the island is where the Portuguese, ever conscious of the need to keep an eye on the seas, would logically have put it, as they built a fortress on Bahrain rather than the mainland of Arabia. Then there was a slab of limestone found in about 1940 and preserved in the Political Agency. Although damaged, its inscription was more or less translatable: "Soteles, the Athenian, and the soldiers . . . to Zeus the Saviour, to Poseidon, and to Artemis the Saviouress." What were an Athenian and his companions doing on Failaka? Shipwrecked voyagers was, until the digging got under way, the most popular explanation.

9. Stone found at Failaka in 1940 and kept in the Political Agency. Its inscription reads: "Soteles, the Athenian, and the soldiers . . . to Zeus the Saviour, to Poseidon, and to Artemis the Saviouress."

The Danes set to work, and in three years they supplied some startling answers to questions not only of Kuwait's early history but to the wider issues of civilisation's first outcrops. A brief dig and a cursory look at the potsherds was enough to show that the fort was not Portuguese at all but an Arab structure with no pre-Islamic connections. The northern township of Qurainiya where the fort stood offered no more fruitful prospect.

By 1958, they had decided to concentrate on the south-east and south-west corners of the island, on the twin mounds at Saad and Said and a cluster of grave mounds near the village of Zor, where the government had provided them with board and lodgings of luxurious proportions by the standard of the usually hard-pressed archaeologist. It was not long before they had found enough of the ridged pottery fragments familiar from the Barbar dig to know, at least to their own satisfaction, that this place and Bahrain had been in contemporary association. They also found many fragments of red brick which were characteristic of Mesopotamian builders, but which, unlike those used by the latter, were unmarked. They searched superficially the nearby east "tell", and found the glazed potsherds of the Greek layers at Bahrain. By 1960, they had excavated and laid out in meticulous detail a Greek temple, as unquestionably Hellenic in its architectural features as the Parthenon. Between the two "tells" a brick-built workshop, complete with kiln, used to make terracotta figures, suggested the third century BC, the Seleucid period, as their date. By the time the excavations had laid bare a town of square formation, there was every indication that this was a thriving, fortified outpost of the Seleucid Empire which, following the death of Alexander, constituted almost the entire land mass from the Nile, across Arabia and Persia to India. Just outside the temple, another article of immense importance was found.

It was a lump of metal which close inspection revealed to be a hoard of silver coins, one of which bore the head of the young Syrian King Antiochus, who from 223 to 187 BC ruled over the Seleucid Empire in the wake of Alexander the Great. The others – twelve in number – had an obverse representation of the Greek hero Herakles, with Zeus on the reverse. They also bore the mail mark of Alexander. Since they were almost certainly buried at the same time as the Antiochus coin – and appear to have been in near mint condition – some hundred years must have elapsed between Alexander's death in 323 BC and their striking.

An explanation of this late minting and appearance at Ikaros of these coins is advanced by Otto Mørkholm, in a paper prepared for the Kuwait Department of Antiquity and Museums. A consequence of Alexander's enormous minting of gold and silver coins, he argues, and of their wide popularity as a medium of payment in international trade, was extensive imitation in many outposts of the Greek world long after their official minting. The coins of Failaka are crude by comparison with original Alexandrian versions. Where, then, were they struck? Similar coins have been known for many years but their place of discovery is not certain, although Celtic tribes of southern Europe often imitated Macedonian coins. In 1951, however, two identical coins were discovered at Gordion in Phrygia, dating from about 210 BC. That find suggests that the barbarianised Alexander coins may have originated a good deal farther east than had been supposed. The Galatians, who migrated from Europe to Asia Minor in the third century BC, have been named as possible culprits. But the deductive processes of the archaeologists indicate an even closer mint.

Of the twelve Alexander coins, eight were struck with the same obverse die, while the same reverse die was used in a number of cases. It is argued that the concentrated use of one or two dies in a single hoard presupposes a direct passage from mint to burial place, without the coins going into circulation meanwhile. Perhaps a more persuasive argument in this connection is the actual condition of the metal, which does not appear to indicate the kind of wear suffered in circulation. It is unlikely that the origin of what amounts to counterfeit money would be found inside the frontiers of the closely ruled Hellnist kingdoms. Certainly not in Ikaros itself since the Seleucid power governed by strict proxy from Greece.

Mr Mørkholm suggests the Gerrhaeans, a prominent trading people of the Arabian mainland as the likeliest source. Their capital, Gerrha, lay somewhere opposite Bahrain, though exactly where, as we shall see, is by no means certain. We know a good deal about their habits from the Roman geographer Strabo who used Hellenic material in charting the demography of this part of the world. He traced their routes along the Euphrates and Tigris to the cities of Susa and Seleucia, beyond to Phoenicia and Syria, and thence to the Aegean. They traded richly, especially in spices, and seemed to have aroused some envy among their neighbours. King Antiochus was constrained to attack Gerrha in 205 BC not so much to gain control over its inhabitants as to secure a fair share of their trade. That accomplished, peace was restored.

This mobile people would have knowledge of Greek versions of Alexander coins, though if they were responsible for minting the Failaka hoard, their die-cutting fell short of faithful imitation.

Before the discovery of this hoard, a few coins had been discovered singly. One was struck by Seleucus I soon after the death of Alexander the Great and is traceable to Susa and Seleucia, while two others belong to the reign of Antiochus III. A small silver drachma, attributed to the Minaeans in the very south of Arabia, was also found. The latter coin is of particular interest since its minting was unknown except for one similar coin in Aberdeen (catalogued by the British Museum). Again it is an Alexander imitation but the usual Greek inscription has been replaced by the name Abyatha – presumably a royal designation – in Sabaean or old Arabian script. Otto Mørkholm suggests a minting date of about 150 BC. The presence of this coin confirms that Kuwait and the small island at the entrace to its bay were on the long trade route from South Arabia to the Mediterranean, just as they were clearly on the caravan path between Gerrha and the Adriatic.

Near the site of the money find, another sizeable clue was unearthed – a rectangular slab of stone containing a long Greek inscription. It consisted of an injunction from a certain Ikadion to the local governor: "The king concerns himself with the island of Ikaros, for his ancestors." It was enough to confirm the evidence of classical writers and historians that there lay to the south of the Alexandrian Empire an island outpost called Ichara or Ikaros near the mouth of the Euphrates; and that the present-day Failaka was that island. The same recorders of the classical period, particularly Strabo and Pliny, referred to another island of the Gulf, Tylos, which was assuredly the place now called Bahrain. Could Tylos be derived from Dilmun, which in the nature of cuneiform script is so easily represented as Tilmun?

There was another means of identification. About AD 170, the Roman historian Arrian wrote an account of Alexander's campaigns based on contemporary records. From this, it seems, the Greek Admiral Nearchos followed Alexander's conquest of Persia with a visit to India in the course of which he sailed up the Indus, now of course in Pakistan, and on the return journey, explored the coastline of Arabia. From this journey, Alexander was informed of two islands. One, nearer to the Euphrates, was thickly wooded and had a shrine to Artemis, and wild goats and antelopes which were sacred to Artemis. Alexander commanded it to be called Ikaros, after the island of the same name in the Aegean.

The shrine was not found by the archaeologists. It was not under or near the temple. Could it have been in the place where, in post-Islamic times, the shrine of Al-Khidr was built? They could not abuse the hospitality of the islanders by digging to find out.

While the east "tell" was divulging its evidence of Seleucid occupation of Kuwait territory, the western mound was also throwing up its share of surprises. The excavators had expected to find steatite seals in the wake of the Barbar sherds that littered the surface. But they were not prepared for the haul that materialised – 290 in all. One bore a face inscription in the untranslated script of the Indus. Apart from this and a few double-sided seals, all were of the third, smaller type found at the fort site in Bahrain. The unity of Failaka and Bahrain, of Ikaros and Tylos, within one of the most ancient known civilisations, was becoming increasingly apparent.

Other objects came out of the west "tell", including fragments of steatite bowls ornamented with figures of men and animals. Among the latter, there was a fragment of two human figures with the cunei-

10. Attempted reconstruction of 44-line inscription on rectangular stone, containing injunction from a Greek, Ikadion, to the local governor: "The king concerns himself with the island of Ikaros, for his ancestors . . .". Much of the inscription remains unintelligible.

11. Reconstruction of Greek temple excavated at Failaka

form inscription: "e-gal In-zak" – the temple of the god Inzak. The archaeologists recalled the first line of the inscription found by Captain Durand in Bahrain, some eighty years before. The bowl from which it came had undoubtedly been the property of an earlier temple dedicated to Inzak, the god of Dilmun. The scholarship with which the archaeologists of the Danish expedition informed its practical activity of digging and sifting was remarkable in its breadth. Here were two sites, close to each other yet separated by 2,000 years. One community had dedicated its temple to Inzak the god of Dilmun, the other to Artemis. And the island remained a holy island, a place to which pilgrims come to pay homage to the Muslim saint Al-Khidr. The holy text of Islam might provide a connecting link, for the Prophet was closer to the events under appraisal.

In the Sura of Al-Furkan (Sura 25, verse 55) they found the words:

"And He it is who hath let loose the two seas, the one sweet, fresh; and the other salt, bitter; and hath put an interspace between them, and a barrier that cannot be passed."

In Sura 18, verse 59:

"Remember when Moses said to his servant, I will not stop till I reach the confluence of the two seas, or for years will I journey on."

In a later verse, Moses met one of God's servants and Moses said to him: "Shall I follow thee that thou teach me, for guidance, of that which thou too hast been taught."

The "two seas" are represented as al-bahrain. Nowhere does the Quran identify al-bahrain as a place. Always it is the "two seas", one fresh the other salt; a perpetuation, surely, of the Babylonian legend of a subterranean sea. But the Recollections of the Prophet assembled after his death, relate that the servant of God whom Moses met at the meeting place of the two seas was Al-Khidr. Al-Khidr it is said had formerly been the vizier of the "One with Two Horns". The Quran explains that Dhu'l Qurnain (He of the Two Horns) travelled to the setting of the sun, and then to the rising of the sun, where he built a wall as a defence against the giants Gog and Magog. It is akin to the Babylonian story of Gilgamesh and his bull-man companion, and it is possible that the One with the Two Horns was Enki and the vizier Ziusudra, whom Enki saved from the Deluge. The story may have changed somewhat from the Sumerian and Babylonian versions to that of the Prophet. But there is enough identity to suggest that the island of Failaka has been a holy island for 4,000 years or more; and that for a large slice of that time it was part of an expanse of the Gulf which stretched for at least 250 miles. The land of Dilmun.

If such a piecing together of archaeological finds and deductive reasoning is to hold together, there must surely be more continuity of geography and history than has been demonstrated so far. It is inconceivable that a common civilisation existed on two islands, 250 miles apart, with nothing in between. There is evidence from the tablets of Assyria and Babylon that Failaka was the northernmost part of such a community. Its proximity to the southern cities of Mesopotamia confirms that likelihood. But what of Bahrain? Was it the southern extremity of Dilmun?

The Danes pursued several cities in their tantalising and ably marshalled effort to achieve in fifteen years as much by way of information, and probably more by way of physical effort, than other expeditions had achieved in twice the time with considerably greater resources. They sought the lost region of Makan, the source of copper in the bronze age, by chasing cultures found at Umm an-Nar, another small island near Abu Dhabi, on to the Arabian mainland at Buraimi and thence to the Sultanate of Muscat and Oman.

They also found another civilisation contemporary with, or perhaps older than, Dilmun; but quite different in its buildings and artifacts. The lives and habits of the people of Umm an-Nar, were more closely related to the early habitations of the Indus Valley than those of the Barbar people of Dilmun. When they went inland to Buraimi, they found yet another culture with signs of Indo-European incursion; Buraimi had close connections with Umm an-Nar, five days' camel journey away, but there were vast differences of custom and buildings. In Oman they found an indigenous civilisation of the third millennium BC, contemporary with Dilmun and the earliest known cultures of Mesopotamia, and this they believe may well be the location of the lost Makan. But political difficulties prevented them from digging in the place they thought most likely to establish the matter beyond doubt. There remained the question of the other, and perhaps most famous, lost territory of Arabia—Gerrha.

Strabo quoted Artemidorus of the first century BC as saying ". . . Gerrhaei have become the richest of all tribes, and possess great quantities of wrought articles in gold and silver". Artemidorus also commented on the magnificence of their houses. Strabo even gave the location of the place, but his directions are not easy to follow. Gerrha was inhabited by Chaldeans from Babylon "about 200 stadia from the sea – you sail onwards from Gerrha to Tylos". The elder Pliny said it was beyond Ikaros and the Gulf of Capeus. He described the city as being five miles in circumference with towers built from blocks of salt. The town of Uqair, a familiar name in the history of Kuwait, seemed to coincide with the positioning given by the Roman chroniclers, who were able to use nearly contemporary Greek sources. But the archaeologists found nothing in the present site to bear out the notion that this was Gerrha. They went across the desert to Thaj and found pottery fragments such as those of Failaka, belonging to the Seleucid period, and Attic ware imported from Greece itself. On some of the buildings of this city, which once stood on a lake, there were inscriptions in Sabaean, the pre-Islamic Arabic tongue of the ancient kingdom of south-west Arabia. On the island of Tarut, opposite Qatif, they found sherds of the Barbar period. Here, at any rate, was an intermediary town of Dilmun, connecting, however tenuously, the distant islands of Failaka and Bahrain with the still incomplete picture of Dilmun. At Qala'at al-Bahrain they went on digging, and found even earlier sherds than the pottery relics of Barbar which they had labelled early Dilmun. They found more seals which, with those of Failaka, suggested that a commercial pulse beat more rythmically between the early cities of the Indus and Dilmun, than between either of those civilisations and Mesopotamia. They found that the weights and measures of Dilmun were those of Harappa and Mohenjo-daro, not of Assyria or Babylonia. The base civilisation of Bahrain, and perhaps of all Dilmun, they were able to date at about 2800 BC, the period of Gilgamesh.

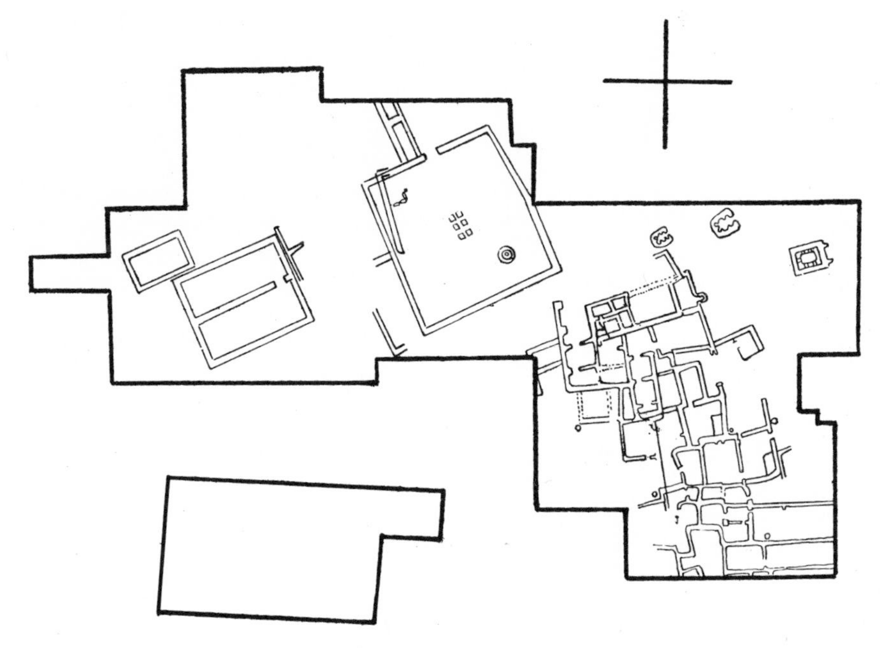

12. Plan of Dilmun city, constructed from excavation of Bronze Age site at Failaka.

By 1968, they were back in Thaj, where they reached the dead end of a city destroyed by fire, and perhaps the sword. They returned to the island of Tarut off the coastal oasis of Qatif. The point at which digging would have been most profitable was occupied by a women's bathing pool and they were not allowed to look upon it except by special dispensation, much less dig it away. Inland in Arabia, they came upon irrigation channels which had become reclaimed land, and potsherds which suggested townships, though there were no buildings. At some time in the past the sea had intruded where now there was desert. Inconsistencies on some of the sites they had investigated suggested that vegetation had grown and fresh water had seeped where now there was arid desert. And from this, they went on to make an interesting speculation. It is generally assumed that during the late miocene age, the last period of mountain formation, the Persian massif lunged and tipped Arabia so that in the east it pressed down below sea level to form the Gulf, while in the west it broke away from Africa to form the Red Sea, with subsidiary cracks in the north at Aqaba and the Jordan Valley. Such, at any rate, is the widely accepted theory. It is, in fact, a theory that was advanced by the Assyrian archaeologists of the nineteenth century. It would certainly explain many of the topographical features of Arabia. But it is to the geologist that one should turn for a more exact account of these matters.

Professor Arthur Holmes, one of the world's greatest geological authorities, has dealt with the matter at some length in his *Principles of Physical Geology*. He concludes his examination of the separation process which produced the Red Sea and the Gulf of Aden as follows:

"The general picture we have so far is of a vast 'swell' (embracing parts of the Nubian Desert and Sinai, Arabia, Somalia and Ethiopia) which has been trisected by the depressions of the Red Sea, the Gulf of Aden and the Ethiopian rift system, all three bounded by faults which have been active at various times from the middle Tertiary to the present day. . . . We have here an excellent example of phenomena which cannot be accounted for by expansion alone or by sub-crustal currents alone. Even with both they cannot be fully accounted for, but both are necessary as minimum requirements."

The issues involved are too specialised and complex for analysis within the context of a history of Kuwait. But the origin of the great land area of which Kuwait is part touches on some fundamental questions of the earth's origin, and for that reason alone it has an endless fascination. For the moment, the common theory provides a likely, and for most of us, intelligible account of the formation of Arabia and the Arabian Gulf. Moreover, the speculation of the archaeologists – based on that theory and arising out of their investigations inland from Qatif – raises another possibility. It is possible, they believe, that the eastern shore which was gradually submerged by the Gulf is rising again. It could be that parts of the region that lay between Failaka, the northern realm of Dilmun and Bahrain in the south, are now beneath the Gulf, and that they will eventually reveal themselves as the coastal shelf re-emerges.

Discoveries of flint and arrow heads, especially at Qatar, have begun to bridge the gap between the Stone Age habitations of the neolithic – even back into the vast darkness of the palaeolithic – and the first civilisations which grew up not, as had been thought, as a crescent from the north-east to the north-west of Arabia, but as a configuration which described a path from the Nile Valley to the Euphrates, along the Gulf and then across Persia to the Indus Valley.

Perhaps future exploration along the mainland of the Gulf, and particularly in Qatar where a large haul of stone implements has already been recovered and where evidence of prehistoric habitation is definite, will enable the map of early occupation to be drawn farther

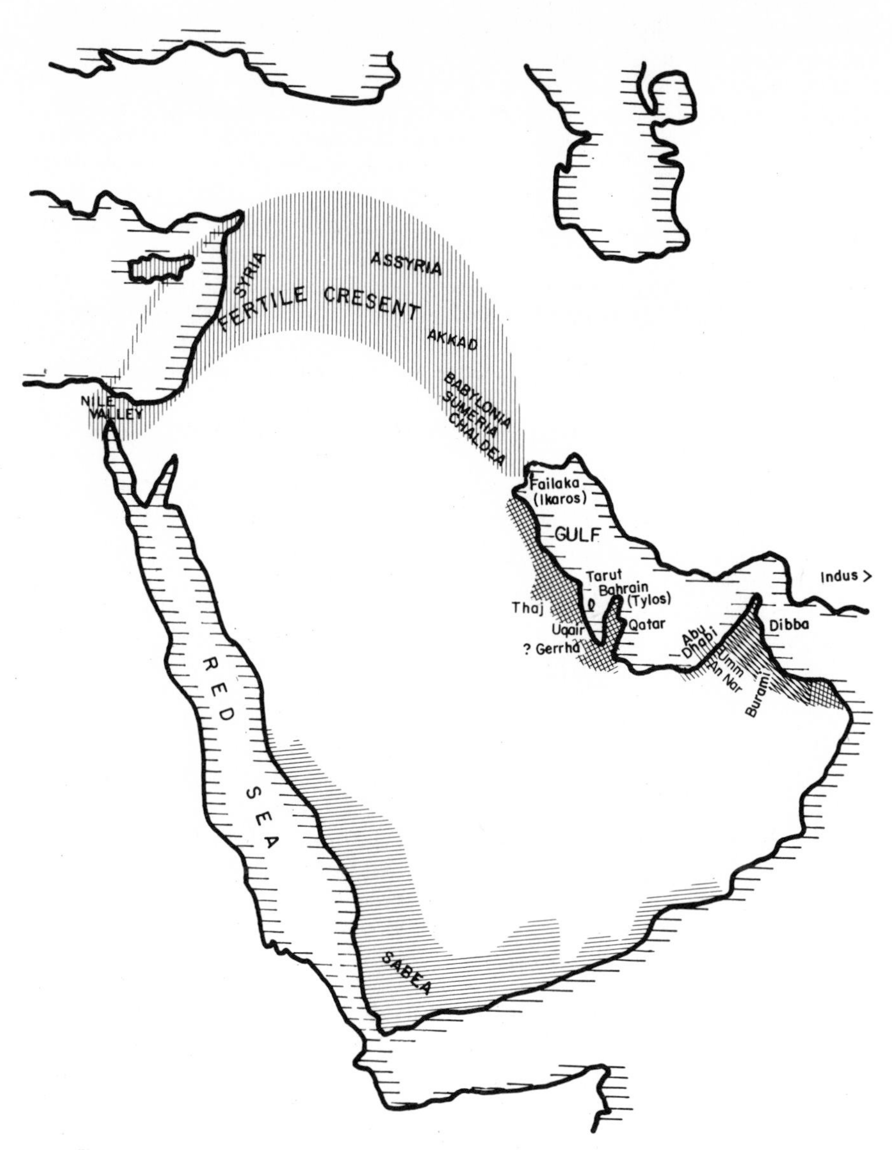

v *Known and conjectural centres of civilisation around the Arabian Peninsula*

back into the shadows of time. The importance of the discovery that the Gulf was at the heart of mankind's earliest civilisation is striking enough. But it is worth recalling in the same breath that the cave art of Cro-magnon man so far revealed ended some 12,000 years ago and stretched back another 10,000 years. Man's emergence through an Ice Age which lasted for a million years and finally melted away a little over 10,000 years ago, is still shrouded in mystery. Further exploration in the Gulf may throw up links in the chain of events which led thence, from the Stone Age to the Copper and Bronze Ages of Dilmun and its northern neighbours.

Isolated villages have been found in parts of Mesopotamia, Iran and Asia Minor which date back to the seventh millennium BC. Jericho, surrounded by its stone wall and rock defences, stood in the Jordan valley 8,000 years BC. But these communities, sometimes substantial, sometimes small and prey to wandering tribesmen, were not necessarily early clusters of civilisation. As Sir Mortimer Wheeler has said, without committing himself to the proposition, "In common usage . . . civilisation is held to imply certain qualities in excess of the attainment at present ascribable to Jericho." Systematic business methods, writing and orderly government are among the qualities we lay down, however imprecisely. Thus, a first assessment of the Dilmun discoveries suggests that here, along the Gulf coast that was al-Bahrain and the islands off it, the earliest civilised communities emerged out of transitional Ubaid cultures, at much the same time as did those of Mesopotamia. And it may be that farther down in Oman, another highly developed trading nation called Makan preceded them all. Much remains to be examined and put to the scrutiny of scholarship. We cannot yet name the political leaders or soldiers of Dilmun as we can those of Babylonia and Assyria. Records of trade with Dilmun have been found in Mesopotamia and perhaps, if ever the Indus script becomes intelligible, similar records will be found in those easterly communities. Yet Dilmun provides no details of its own except, of course, for the material evidence of imported goods and the hoards of stamp seals and pottery which are in themselves indicative of trade. Drainage, town planning and municipal zeal seem to have been as much part of Dilmun's make-up as of Babylon's or Mohenjo-daro's. But art and craftsmanship, if seals and pottery of the Barbar culture are to be compared with similar manifestations of those other cultures, look at a glance to have been less developed. Some of the Indus seals are by any standard miniature works of art. Dilmun's seldom look more than workmanlike.

The substance of the Dilmun discovery is nonetheless of major importance. As important perhaps as the first finds of Mesopotamia which took the best part of fifty years to make any real impact on the world at large and almost a century to reach their culmination in the vast discovery of Ur. In looking at the achievement of Professor Glob and Geoffrey Bibby and their colleagues in the Gulf, it would be hard to set down a better conclusion than that written by Sir Henry Layard, the digger of Nimrud: ". . . a deep mystery hangs over Assyria, Babylon and Chaldea. With these names are linked great nations and great cities dimly shadowed forth in history, mighty ruins in the midst of deserts, defying, by their very desolation and lack of definite form, the description of the traveller." Had he known, he might well have included Dilmun.

3. The Islamic Heritage

Alexander's empire came to an end in 323 BC with the death of the great Macedonian at Babylon. His provinces in Persia and the Arabian peninsula were taken over by the Seleucids from the region that was once Assyria. They held sway until 226 BC. Contemporary writers and cartographers marked the bay of Kuwait as the Gulf terminal of the empire. In the third century AD, the Sassanids brought a return of Persian rule, which lasted until the year 642.

If the excavations on the Gulf islands and coasts are only now beginning to throw light on the ancient cities and trade of the eastern seaboard of Arabia, there was one aspect of life on the peninsula which went on largely unchanged through several millennia to comparatively recent years; in that expanse of time the nomadic bedu of the Arabian steppes kept up a way of life of almost incredible antiquity. According to the archaeologists, animals were domesticated in the Near East about 10,000 BC, and as Dr Glyn Daniel has written in *The Idea of Prehistory*: "from the single basis of domestication came into existence the nomadic societies of Arabia, North Africa and Central Asia." The nomads existed from prehistory, they were already there when man began to leave evidence of the civilisations of Babylonia, Dilmun and Egypt.

The settled populations of the oases and river valleys were always an easy prey for the attacks of these fierce desert men, though at times the overspill of desert people would themselves settle on the land, bringing fresh, vigorous blood to the village communities. Clay tablets from Mesopotamia record incursions by desert nomads from earliest times, and much later, in 650 BC, the subject reappears in a carved relief in the palace of Assurbanipal at Nineveh, which shows an attack by camel-borne invaders from the desert. The Old Testament's portrayal of Abraham's way of life, and references to the black tents of Kedar, are an indication that even as far back as the second millennium BC, the bedu lived in a way which scarcely differed from that observed by Europeans in Arabia in the early years of this century.

There is one other society of pre-Islamic Arabia whose existence is also mentioned in the Old Testament – that of the South Arabian Kingdoms. The early Kingdom of Yemen had been succeeded by a group of smaller states of which Saba (the Sheba of the Old Testament) was the most prominent between 750 and 115 BC.

It was the Sabaeans who developed the important and lucrative carrying trade in frankincense, myrrh and gold from their own lands and Ethiopia; spices, and jewels from India and China; pearls from the Gulf. From about 400 BC, the neighbouring kingdoms of Main, Qataban and Hadramaut shared in Saba's prosperity. For several centuries their caravans plied up the "incense road" along the west coast of Arabia, passing through Mecca to Egypt, Gaza and Damascus, whence their goods found their way to Rome. At the peak of their wealth and power it seems likely that these South Arabian states had colonies in Oman and North Arabia. But in the third century BC, their rich trade began to decline as Egyptian and Roman ships from the Red Sea started to make their own voyages to the orient.

In the third or second century BC, Saba and the neighbouring kingdoms all came under the control of the Himyarites, whose power lasted 600 years until AD 525.

Among the prosperous trading communities of the south-west corner of the peninsula, the South Arabian script was developed. The first examples of inscriptions on stone in these characters were discovered in the Yemen in the 1830s, and it was at first thought that this script (sometimes incorrectly called Himyaritic) was peculiar to that region. Since then further examples have been found as far afield as Ur and Palestine, as well as from many places in Saudi Arabia. Many examples ascribed to the period AD 200–500 have come from Thaj, site of an ancient city in Hasa about 120 miles south of Kuwait and one of the places examined by the Danish archaeologists. Captain Shakespear, while Political Agent in Kuwait, was the first to find South Arabian inscriptions in the Gulf area when he made two discoveries at Thaj in 1911. These inscriptions – most of which are grave memorials – are written in varying forms of Semitic language, closely related to Arabic, and it is now believed that this pre-Islamic script was used throughout the period from 800 BC to AD 450 to write not only the language of Saba and Main but also the Thamudean and Lihyanite dialects of the area north of Saba and the Safaitic speech of the eastern Syrian desert.

The Sabaeans were known to have been worshippers of a moon-god, as also were the Himyarites who succeeded them. At this time Mecca was already a holy city, indeed its name originated in a Sabaean word for "sanctuary". The sacred stone of the Kaaba was supposed by the moon-worshippers to have been dropped to earth by their god and the thriving city which lay on the main caravan route from south-west Arabia had, long before Islam, become doubly prosperous as a centre of pilgrimage.

Many other primitive religions of the time were connected with fertility symbols such as water and trees. The well of Zam-zam which was believed to have succoured Hagar and Ishmael, and which stood in the precincts of the Kaaba, was also hallowed before Islam. The sanctuary at Mecca was filled with hundreds of idols. And, like many flourishing commercial centres, Mecca itself became a city of corruption and immorality. This was the city into which Muhammad, the founder of the Muslim faith, was born in AD 571.

The prophet Muhammad called the people of Mecca to turn away from their pagan superstitions and belief in multiple spirits, and to reform their immoral and disordered lives. The religion which he founded therefore places great emphasis on moral behaviour, laying down strict rules with regard to almost every aspect of life. In its theological aspect it is a religion of stark simplicity, without doctrines. Its adherents hold to the simple principle of the unity of God and his universality, summed up in the phrase "There is no God but God and Muhammad is his prophet".

The five obligations of the good Muslim are:

To accept the one God and his prophet Muhammad;
to pray five times a day;
to give alms;
to keep the fast of Ramadhan;
to make the pilgrimage to Mecca.

Islam is a religion without priests; it requires no mediation between God and man. The imam of a mosque is a leader noted for his piety, and therefore chosen to lead the prayers. Friday is the Muslim Sabbath, and in most mosques there is special teaching by the *khatib* or preacher on that day, so that though a man may pray anywhere at other times, attendance at a mosque is virtually obligatory on a Friday.

The Quran, which contains the truths revealed to Muhammad by God, describes the attributes of God and his relationship to man, and lists the religious and moral obligations binding upon those who would submit themselves to God's will. Among its most important teachings are those relating to marriage and divorce, the position of women and children – particularly with regard to widows and orphans – the division of property, the treatment of slaves, warfare and sharing the spoils, and alms-giving. While the tone of the Quran is often one of righteous indignation, it is also full of human understanding, prescribing forgiveness to the penitent, kindness to the unfortunate, practical charity to the poor, and patience in all things.

The teaching of the Quran is the basis of all Arab culture. Not only has it been disseminated by scholars and divines through the ages, but the lilting rhythms of its poetry enable it to be easily committed to memory. In spite of its many obscurities, it has thus become part of the oral tradition of generations of illiterate people. Among townspeople many men could recite the whole Quran by heart, and even among the bedu many of its pithy sayings and more memorable portions were part of their common heritage of knowledge. Many of the prophet's practical injunctions were framed to suit the particular conditions of life in Arabia in the seventh century, but other rules have a more important and general application which is valid for all time. The firm belief in one God, to whom every man is individually answerable for his conduct, and the promise of appropriate rewards or punishment after death, make the true Muslim a man of noble principles and high moral standards. But like other religions, Islam has on occasion fostered movements in which the interpretation of the faith has been carried to extremes of bigotry and intolerance.

Kuwait before the days of oil was a Muslim state barely touched by outside influences. The original population were all of the Sunni sect, but there was also a large Shia group, many of them of Iraqi or Persian origin. Together they formed a God-fearing community for whom religion coloured practically every thought and action of their lives, in a way that has something of a parallel in medieval Europe. The observance of the five daily prayers was never a meaningless formality;

it was the homage paid to a watchful ever-present God who could punish those who flouted his laws. Similarly, the fast of Ramadhan was an exercise in self-denial by which men proved their personal devotion. Because of the lunar calendar the month of Ramadhan occurs at varying times by the western calendar (each year it falls a little earlier than in the previous western year). This means that there will be times when the period of fast has to be observed in the heat of summer. Since the August shade temperature averages 112°F, and even a drink of water is forbidden in the hours of daylight, the fast at such times is a severe hardship. During a summer Ramadhan in old Kuwait most people used to sleep as much as possible during the day, and when thirst became almost intolerable a man would rinse his mouth with water and spit it out.

There were few comforts in the average Kuwaiti family's daily life in those days, and no protection against sickness or epidemics. For both bedu and townspeople life was full of uncertainty and in the natural hazards of the wilderness and the ocean all were conscious of man's helplessness and the need for a protecting God. Favourable conditions – good rainfall and pasture, or fair sailing-winds and a good trading profit, or the natural joys of family life, such as the birth of children – were always seen as manifestations of God's bounty. Gratitude for his beneficence was constantly on their lips, but complaint never. When disaster or distress came their way they accepted it fatalistically, especially the bedu, whose control over their emotions was absolute. They believed that though Allah had a propensity to mercy he had foreordained the events of their lives so that it was pointless to bewail misfortunes, or indeed to take precautions to avoid trouble. If a European expressed regret at someone's affliction or hardship, the answer was nearly always "*Maktub* – it is written". Among the townspeople, however, emotions were not always so strictly held in check, and in times of bereavement women would rend their clothes or give violent and hysterical expression to their grief.

Theft, dishonesty, or crimes of violence were rare in old Kuwait town and when such offences occurred in the desert there were well-established rules for compensation in money or blood. There can have been few societies where the accepted moral code was more universally observed, and it was recognised by everybody that the code was God's law rather than man's. An evil-doer was by definition "one who does not fear God" or "one who neither prays nor fasts".

The pilgrimage to Mecca was everybody's ultimate ambition, and many of those who made it spoke of it afterwards in a way that showed that it had been a genuine source of strength and inspiration.

Today, in the changed material conditions of Kuwait, religion no longer exerts such a strong influence over men's lives, and the individual's observance of prayer and fasting is often less strict than it used to be. More sophisticated Arabs will even argue that some of the Islamic rules of life, such as abstinence from alcohol, are no longer relevant in a changed society. Nevertheless, Kuwait remains a Muslim country, and a natural unquestioning belief in God and in the teaching of the Quran remains deep-rooted in the Kuwaiti character, a comforting and stabilising influence through times of bewildering change.

But we must return to the growth of Islam from its early beginning. The prophet Muhammad lived from AD 571 to 632, and by the time of his death his new religion was well established in the western province of Arabia around Mecca and Medina, and accepted by the neighbouring tribes in Najd and down to the border of Yemen.

In the period immediately following the prophet's death, the conquering armies of Islam carried the faith outwards from its original nucleus. First its domain was extended eastwards across the peninsula, and then northwards into Mesopotamia and Syria.

The population of infertile Arabia had always overflowed to the fertile lands of Syria and the Euphrates valley, and the Muslim armies which swept north and east into those countries were probably impelled by economic pressure as well as fanaticism to follow an ancient pattern of migration, though this time they were inspired by the belief that to acquire land and booty from infidels was morally just, and that any who died fighting for Islam against pagans went straight to paradise.

The famous soldier Khalid ibn al Walid (an important thoroughfare in modern Kuwait bears his name), commanded the Muslim army which after Muhammad's death, during the caliphate of his immediate successor Abu Bakr, embarked on the eastward campaign of conquest. At this period the Sassanids ruled Mesopotamia and Persia and also dominated much of the Arabian littoral of the Gulf, including Bahrain, Oman and areas as far south as the Hadramaut. In 632 or soon after, Khalid ibn al Walid led his armies, reputed to have numbered 20,000 men, in an attack on the western province of Persia. His tribal followers, not only fired with zeal for the new faith but magnetised by the personal prowess of Khalid "the Sword of Islam", proceeded north-eastwards through what would today be Kuwait territory, and Khalid joined battle with Hormuz, the Persian governor of the province, at a place which the old Arab historians name as Ubulla. It was a point at the head of the Gulf near the river which is today known as the Shatt al Arab. The exact site of Ubulla was probably near Zubair, though some believe it was at Kadhima, on the north shore of Kuwait bay. But if the actual location is uncertain, there is no doubt that even in antiquity the north-west corner of the Gulf was a strategic point, for here the ancient caravan trails from Syria and Najd came to the sea and met the trade routes from Persia.

Before the Arab and Persian forces joined battle, Hormuz the Persian commander, anticipating an easy victory, is said to have prepared fetters for his Arab prisoners. Because of this the battle which followed has become known as the "Battle of the Chains", but the over-confidence of Hormuz proved to be ill-founded, for he and his armies were roundly defeated and the victorious Arabs relieved them of vast quantities of booty.

During the two years of Abu Bakr's caliphate the whole of Arabia achieved real unity under the banner of Islam. Shortly afterwards, however, Islam was split by disagreement over the succession to the caliphate. The conflict over this issue was to lead finally to the division of Islam into its two major sects which later came to be called the Sunni and the Shia.

The Sunni sect, to which practically all of Arabia proper (including Kuwait) adheres, originated in that section of Islam which after Muhammad's death accepted the caliphate of Abu Bakr and his nominated successor Omar. However, Ali, the prophet's cousin and husband of his daughter Fatima, was regarded by a large faction as the rightful successor, and their discontentment grew as they saw Ali's claim set aside through three successive caliphates. When Othman, the third caliph, was murdered Ali's supporters seized their chance and proclaimed him caliph, while those who refused to accept him rallied behind their own leader Muawiyah. Since opposition to him was most concentrated in Syria and Iraq, Ali went to Iraq, where his men were defeated in battle by forces rallied against him by Ayesha, the prophet's favourite wife, and by Muawiyah. Once Muawiyah had proved himself the stronger leader, he was proclaimed caliph in Jerusalem, becoming the first of the Omayyad dynasty. Shortly afterwards Ali was assassinated outside the Kufa mosque.

The Shia sect sprang from those who believed Ali was the true caliph, but the most distinctive feature of their religion is the homage paid not only to Ali but to his sons Hassan and Hussain, particularly the latter, whose martyrdom annually provokes ceremonies of frenzied grief throughout the Shia world on the tenth day of the month of Muharram. The eldest son Hassan was murdered by one of his wives in Medina eight years after his father's death. The younger, Hussain, returned to Iraq to lead a new fight against the Omayyads, now led by the caliph Yazid. But Hussain had been misled as to the comparative strength of his own and Yazid's followers, and when he met a force of 4,000 of Yazid's men he and his small band were almost surrounded and totally outnumbered. Knowing they faced certain death but convinced of the justice of their cause, Hussain and his men spent the night before the impending battle digging a trench which they filled with burning brushwood to prevent their own retreat. The following morning Hussain's force was totally destroyed. He himself was the last to die, cut to pieces and trampled upon by the Omayyad forces. His head was removed and taken to Kufa.

For the Shia, Ali, Hassan, Hussain, and their nine direct descendants are the only true caliphs. The mosque of Najaf in Iraq represents the tomb of Ali, and that at Karbala is the shrine of Hussain. Both places are important centres of pilgrimage for the Shia world.

The desert tribes who rallied to the leadership of Khalid ibn al Walid belonged to those aboriginals of Arabia who had been shaped by their habitat into one of the most perfectly-adapted of primitive human societies. They were tent-dwellers, owning camels and sheep, who moved constantly in search of pasture, a proud people, courageous in battle, toughened by the constant struggle for survival.

The beduin Arab has been the subject of much romanticised treatment by English writers of fact and fiction. Austere Victorian travellers eulogised his way of life, hardened soldiers and explorers waxed poetic about his virtues. His appeal for all such Europeans lay chiefly in the fact that in all things he was his own master. He was free of political control, unfettered by house or material ties, unrestricted in his wanderings. He deferred to no man, and was subject to no man's bidding.

Gibbon's gift for grasping the essentials of a subject has enabled him to sum up the character of the desert Arab with admirable force and clarity:

"Their spirit is free, their steps are unconfined, the desert is open, and the tribes and families are held together by a mutual and voluntary compact. . . . The nation is free, because each of her sons disdains a base submission to the will of a master. His breast is fortified with the austere virtues of courage, patience and sobriety; the love of independence prompts him to exercise the habits of self-command; and the fear of dishonour guards him from the meaner apprehension of pain, of danger, and of death. The gravity and firmness of the mind is conspicuous in his outward demeanour: his speech is slow, weighty and concise; he is seldom provoked to laughter . . . and the sense of his own importance teaches him to accost his equals without levity and his superiors without awe."

Eighty years later, Sir Richard Burton, with close personal knowledge of the bedu, saw in their freedom the essence of manliness for which men in more complex societies, hedged about with prohibitions, secretly yearn. "There *is* degradation, moral and physical, in handiwork compared with the freedom of the desert. The loom and the file do not conserve courtesy and chivalry like the sword and the spear."

By contrast, W. G. Palgrave, among the foremost Arabian traveller-writers of the nineteenth century, took a more disenchanted view of the Arabs, and found the ferocity and bigotry of many of them intolerable. While recognising the strength and determination of the Arab character he considered these more admirable qualities were too often mixed with guile. His judgement of the people of Aridh – the native province of the Al Saud dynasty – shows an almost prophetic insight in the light of later events by which Abdul Aziz ibn Saud returned to power in Arabia. Indeed, his summing up of the Najdi character could just as well be applied to the famous Shaikh Mubarak of Kuwait: "great courage, endurance, persistence of purpose, an inflexible will united to a most flexible cunning, passions that can bide their time, and audacity long postponed till the moment to strike once and once only."

Most of the nomads to be found in the Kuwait hinterland up till the 1940s belonged to the great tribes of east and central Arabia. They were uncompromisingly beduin, untouched by foreign ideas, unfamiliar with any world except their own. The men were spare and wiry in physique, with strong aquiline noses set in brown faces, and hair hanging in plaits from under their *kaffiyas*; the elders of a tribe were bearded like Old Testament prophets. Their clothing was shabby, never washed and rarely changed, but their bearing was proud, their movements dignified, and they faced the world with a firm and penetrating gaze, conscious of their superiority. Their womenfolk, who could usually be seen busy around the kitchen-quarters of a camp, were clothed in long coloured gowns, their heads draped in black, and their faces covered with the *burga* or mask with two eyeholes. They remained veiled even when attending to the cooking or household tasks. At the approach of strangers they would usually withdraw inside the tent if their menfolk were at home, but if they were alone and the visitors were Arab, their code permitted them to offer rest and shelter to male travellers.

The tents in which they lived were black, woven by the women from the wool of their own sheep and goats; long and low in shape, they were divided into separate sections for men and women. The men's section was immediately recognisable by the coffee-pots which stood ready by the brushwood fire, while the women's section would often be more enclosed by having part of the side of the tent brought across the front to give some privacy. There were always children round a camp, but the infant mortality was high, and probably nearly 50 per cent of all beduin children died before the age of six. Those that survived were the toughest and hardiest of the breed, and by this ruthless process of natural selection the continuity of the race was ensured.

Most of the bedu were poor by European standards, and their possessions were limited to the simplest necessities: a tent, a few rugs to spread on the desert sand, cotton quilts to cover themselves at night, a camel-saddle for the master and a cane-framed *maksar* or camel-litter for the wife, water-skins, utensils for cooking and serving food, and the boxes or saddle-bags in which these things were packed when they moved camp. In addition every man, unless desperately poor, had a rifle and ammunition, the latter usually slotted in a cartridge-belt or bandolier.

The need for mobility in their lives means that property is reduced to the minimum, and in the desert economy wealth is accumulated in livestock. For the average beduin of a good tribe this might be half a dozen camels and thirty sheep, but for a tribal chieftain and his family several hundred camels. The great *sharif* or noble tribes of Arabia invest almost exclusively in camels, keeping sheep only as an immediate food supply. For such tribes the care of sheep is undignified, and they will employ a separate shepherd tribe of lower standing to look after their flocks. The camel above all is vital to the desert life, the indispensable animal half of the man and beast partnership. The camel needs man to draw water for it from the desert wells, and to guide it to new pastures when its own area is grazed bare. Man needs the camel to carry his tents and gear when he moves; he needs the milk of the female camels as the most important single item in his diet; he needs camel dung for fuel. The camel carries the nomad into battle, takes him to buy supplies in town, bears him on a thousand-mile journey across the peninsula to the House of God at Mecca. Finally, though camels are not kept principally to supply meat, they may do so when the need arises. Though the bedu could never put such an idea into words, the camel is not only a practical necessity, but a powerful symbol of status, of security, of mobility, and as such can have an almost mystical significance. The welfare of his camels is every tribesman's first consideration. The acquisition of camels is the usual purpose of beduin warfare, and the loss of them in defeat a bitter humiliation.

The rigours of desert life have dictated the importance of hospitality and other customs which ensure a man's safety in the harshest of natural conditions. The relationship between tent-owner and guest has an almost religious sanctity, and a similar bond and obligation exists between a man and his *khowi* or travelling companion. The custom of *dakhala* or sanctuary entitles a man pursued by an enemy to claim the protection of a third party by uttering a time-honoured formula which has always been regarded as totally binding. A chivalrous attitude to women in warfare has generally been faithfully observed except at moments of blind fanaticism.

Lacking material wealth, the bedu came to prize qualities of character. Concepts of honour, loyalty, dependability, bravery and obedience to religious and social laws were of paramount importance to them, often dearer than life itself. However, fighting between rival tribes, and raiding enemy tribes to steal camels and sheep, was not considered dishonourable; this had always been one of their traditional pursuits, regarded as an adventurous game in which young men might test their qualities of initiative and courage, a sport rather than serious warfare.

The nomadic life shaped the tribal structure of beduin society, and the only loyalty a beduin knew was to his tribe, except at those recurring periods of Arab history when the personality of a dominant leader welded groups of tribes together in a common cause. Mutual dependence, often strengthened by belief in descent from a common ancestor bound a tribe together, and from this solidarity arose such practices as blood revenge for killing.

From earliest times every beduin tribe has had its *dira* or home territory, across which it roams in winter and spring, and where there are wells which it uses in summer. As long as a *dira* is sufficient for its needs a tribe does not try to extend it, but friendly agreements with neighbouring tribes often enable it to use grazing grounds beyond the confines of its own recognised area.

The main tribes whose traditional grazing areas lie around and extend into Kuwait are:

Mutair. *Dira* immediately south west of Kuwait; centred on Irtawiyah.
Awazim. East and south-east down the coast.
Ajman. East and south-east down the coast.

vi Tribal Arabia, as drawn by H. R. P. Dickson. Boundaries are approximate

Sbei and Sahul. South towards Riyadh.
Harb and Shammar. To the west.
Dhafir and Muntafiq. North-west and north.

The modern state boundaries of Kuwait often cut across old allegiances. For example, Kuwait territory has long been part of the Mutair tribal *dira*, and from as far back as they can remember this tribe had acknowledged the authority of the Sabah until, on the death of Shaikh Mubarak, Ibn Saud enticed them to switch their allegiance to himself. When the Mutair rebelled against Ibn Saud from 1928–9 and sought refuge in Kuwait they pleaded – without avail – to be allowed to return to their traditional loyalty to the Sabah. Another tribe, the Awazim, occupy the territory which stretches down the coast south of Kuwait, now lying in Saudi Arabia, though many of them consider the ruler of Kuwait their true overlord. Some tribes have formed confederations, allowing wider grazing-lands to be used by members of the group. Such tribes will assist each other in warfare, or against attack by a rival group. Examples of these confederations are the Harb, Mutair and Ajman (this grouping gives the tribes control of the whole of north Arabia); Bani Abdilla, Mutair and Ataiba; Dhafir, Shammar and Awazim; Ajman, Murra, and tribes from Najran (a blood federation).

Each tribe normally maintains a trading connection with one particular town. Kuwait has traditionally been the supply town for the Mutair, Harb, Shammar, Awazim and northern Ajman. So long as relations between neighbour states have been good, bedu from Saudi Arabia and Iraq have always entered to graze in Kuwait territory or vice versa, for no political boundaries will check the nomad in his migrations to better pastures in the area which he regards as his *dira*. Among smaller tribes, the Rashaida were a group which belonged specially to Kuwait. They were the traditional retainers of the shaikhs of Kuwait, and were noted for the beauty of their women and the hunting skills of their men.

Though each tribe acknowledges one paramount shaikh or chief, such leaders are regarded as no more than individuals chosen by their peers. The desert favours equality among men. The position of a tribal shaikh, while remaining in the same family, does not pass necessarily from father to son. In each generation the aspirant to the chieftaincy must have shown that he has courage, leadership, and the essential quality of good luck. A shaikh cannot give orders to his tribe – the beduin character revolts against direct commands, but after holding a tribal council a leader who has the right qualities of character will carry opinion with him.

The leading family from which a tribe chooses its chief may in some cases be so numerous as to form a clan of its own within the larger tribe. In addition there may be other subsections in the tribe, each having its own lesser shaikh who acknowledges the overall leadership of the paramount chief.

A fairly typical example of tribal structure may be found in the Mutair, the powerful sharif tribe which has played a dramatic part in the history of Arabia in the last fifty years. The tribe divides into three main groups: Al Dushan (the clan which is in effect the tribe's ruling family), Al Ilwah and Al Buraih.

The ruling clan has no subdivisions, but the other two clans within the tribe divide again; the Buraih are composed of Aulad Wasil (in the west) and Aulad Ali (in the east), and Bani Abdilla. The Ilwah subdivide into Al Muwaha, Al Thawiaun, and Al Jiblan. Each of these six subtribes then divides again into sections, each having its recognised leader. There may be as many as ten subsections in each clan within the tribe, each having anything from thirty to a hundred tents. The paramount shaikh of the tribe today is Bandar bin Faisal al Duwish and fifteen years ago it was estimated that the fighting strength of the tribe was in the region of 6,000 men.

Arabs of desert and coast

The people of the coastal townships of the Gulf have always been different in character from the true desert man. Though the early colonists were nomads, once settled their interests took a new direction as their lives were modified, and racially they became more mixed as they intermingled with Persians who had long been settled in scattered groups along the Arabian coast, and with negro slaves who were always present in sea-ports.

After the establishment of the Kuwait settlement under the Sabah family, the natural harbour there and the town's position as the obvious trading post for interior Najd and the Qasim must soon have turned the interests of the Kuwaitis to sea-faring. Arabs had been sailors from earliest times, and long before Kuwait appeared on the map Basra had been a prosperous port and commercial centre, engaged in the trade between the Mediterranean and India. But Kuwait, with its special advantages of a sheltered bay, and a long, gently-shelving foreshore where boats could be built, soon began to rival Basra as a sea-port. By the end of the eighteenth century its shipwrights and sailors were reputed to be the most skilled in the Gulf.

The development of the intense maritime activity upon which Niebuhr later commented must have depended upon an early influx of Baharina people, for this special group have always been the master shipwrights of the Gulf. They are a Shia people, believed to have come to Kuwait from Bahrain. They have large communities both there and in Hasa. Their origins are obscure, and though their religion suggests that they may originally have come across from Persia, some historians believe that they are pure Arabs, converted to the Shia sect at some unrecorded time in the past. Within their close-knit community the traditional skills of boat-building were passed from father to son, and the Baharina had a virtual monopoly in this trade.

Thus there grew up in Kuwait that duality of interest and outlook which gave the old town its distinctive and attractive character. The shaikhs and the founding families had been bedu, maintaining blood connections with the desert; though they ruled a community of merchant venturers, sailors and pearl-divers. Here the desert met the sea, and here bedu and seafarers mingled, while the shrewd Kuwaiti merchants and shopkeepers occupied the middle ground, making their livelihood by exploiting such opportunities as were offered.

For such a community to prosper, a balance had to be maintained between the interests of desert and town, and this required a strong shaikh who could act as a unifying and directing force. In the interests of the merchants the shaikh had to command the good will of the surrounding tribes, or the trading caravans of the town were at their mercy. But the tribes were also dependent upon the good will of the shaikh, who could deny them access to essential water-wells, or prevent them entering the town to buy supplies. More often it was in the interests of both sections to reach a *modus vivendi*, and a town leader would win over a tribe by lavish gifts to its chiefs, who would then acknowledge his authority.

Rise of the Wahhabis

During the centuries which followed the first triumphant extension of Islam, many of the inhabitants of north and central Arabia had gradually grown away from the faith, and pagan practices and superstitions which had been current in the "days of ignorance" had reappeared among them. Even those who still professed to be Muslims had to some extent ceased to observe the stricter disciplines of their religion.

Yet the streak of violent puritanism in the character of the nomads of central Arabia persisted. They led hard lives, devoid of material luxuries, an existence stripped to the bare essentials, in which they were proud of their own hardiness, scornful of the luxury of softer men in gentler societies. For the bedu it was but a step from pride in their way of life to finding a special virtue in their own asceticism, and denouncing neighbouring peoples for self-indulgence and excess. It only needed the emergence of a powerful religious leader to awaken these latent forces. Did not the Quran itself enjoin self-denial and the abandonment of superstitious abuses?

Such a leader appeared during the eighteenth century in the person of Muhammad ibn Abdul Wahhab (1691–1787), who preached a return to the true original teaching of the Prophet. His austere puritanism was so extreme that many orthodox Muslims considered it a heresy more than a reforming movement, and his followers came to be known as Wahhabis among those who would not dignify them with the name of Muslims. The new teaching prohibited alcohol, tobacco, silk and gold in dress, games of chance, magic, and monuments or tombstones for the dead. In particular Muhammad ibn Abdul Wahhab attacked the Hanafi form of Islam as practised in Turkey, and the Shia beliefs of Iraq and Persia, whose customs were abhorrent to the pure Sunni tradition. To the converted Wahhabis, anyone who was not of their persuasion was an "enemy of God" and to kill him and plunder his goods was not only lawful but meritorious. They believed in the Quran as a complete and adequate guide, and anything not specifically commended therein was regarded as forbidden.

In Najd, whose people were most isolated from outside influences, and where the conditions of beduin life were most primitive, the new creed was embraced with unbridled fervour, and in the middle of the eighteenth century a militant fanaticism ran through the bedu of this region like a spark through stubble. Muhammad ibn Saud, a chief of Aridh, one of the Najdi provinces, and an ancestor of the royal Saudi house, was converted about 1742, and was recognised among his own people as the head of the reformed religion, becoming the first Wahhabi amir of Najd in 1745. Under him the movement took on a national as well as religious aspect, and he led his tribal followers in a series of campaigns in which the driving force of their new zeal carried them to one victory after another. Though Muhammad died in 1765, his son Abdul Aziz succeeded him and continued in warfare against the surrounding areas. One of the places where the new creed took hold most strongly was on the coast of Oman among the Qawasim tribes, whose traditional pirate activities were given new justification by Wahhabi injunctions to attack and plunder infidels. But outside Najd there were many who hated the fierce bigotry of the Wahhabis and resented the religious discipline they sought to impose. Among such, "men who would gladly see more tobacco and fewer prayers", to use Palgrave's descriptive phrase, were the Bani Khalid of Hasa and Qatif, who had profitable trading connections with Persian Shias; they were forced to pay tribute to the Najdi amir, though never really converted. Others who opposed Wahhabiism were the northern tribes of the Dhafir and the Muntafiq, who were different in character from the bedu of Arabia proper.

At that time, Shias, regarded as infidels by the Wahhabis, had been forbidden passage through Najd on the traditional pilgrim route from Hasa to Mecca. This had aroused such animosity in the Shia world that the Amir Abdul Aziz was assassinated by a Persian from Karbala in or about the year 1800.

During the rise of this fanaticism Kuwait remained on the periphery of main Wahhabi activities, but as an integral geographical part of Arabia it could not fail to feel the effects of the general upheaval in Najd. Though Kuwait, like Najd, was predominantly Sunni, it traded with Shias in Persia and Iraq, and already had a fair number of Shia inhabitants. The tolerant attitudes to outsiders which naturally develop in a trading settlement were not conducive to a ready acceptance of the Wahhabi creed. It may be that in the interests of self-preservation Shaikh Abdullah I of Kuwait made some tentative pact with the Wahhabis; if so, it did not save his territory from attack. Several attempts were made to incorporate Kuwait in Saudi domains, but at no time was it subdued. Indeed, during the years 1775–9 Kuwait was enjoying special prosperity, because much trade had been diverted through the town during the siege and occupation of Basra by the Persians. Clearly the merchants of Kuwait were at this time preoccupied with commerce rather than with religious wars. Among the bedu in the surrounding desert, however, many would have joined the Wahhabis, in religious conviction if not in actual warfare.

The successful Wahhabi armies invaded Mesopotamia in the 1790s, provoking Turkish retaliation in the form of an expeditionary force sent from Baghdad to Hasa in 1798. The Turks' failure to capture Hufuf, and their final retreat, gave a strong boost to Wahhabi prestige, and offers of submission came in from tribal leaders who had hitherto held out against Najdi domination.

By 1800, the Wahhabi state, based on its capital at Dariyah, had a regular government with a centralised administration. The power of the Al Saud was such that under Saud, the warrior son of Abdul Aziz, they could raise an army of fanatics powerful enough to sack the Shia shrine at Karbala in 1801. Wahhabi power reached its peak in 1803 when Saud and his armies subdued Taif and entered Mecca. By 1808 they had moved northwards and were at the gates of Damascus.

The advances of the Wahhabi armies which seemed almost to be

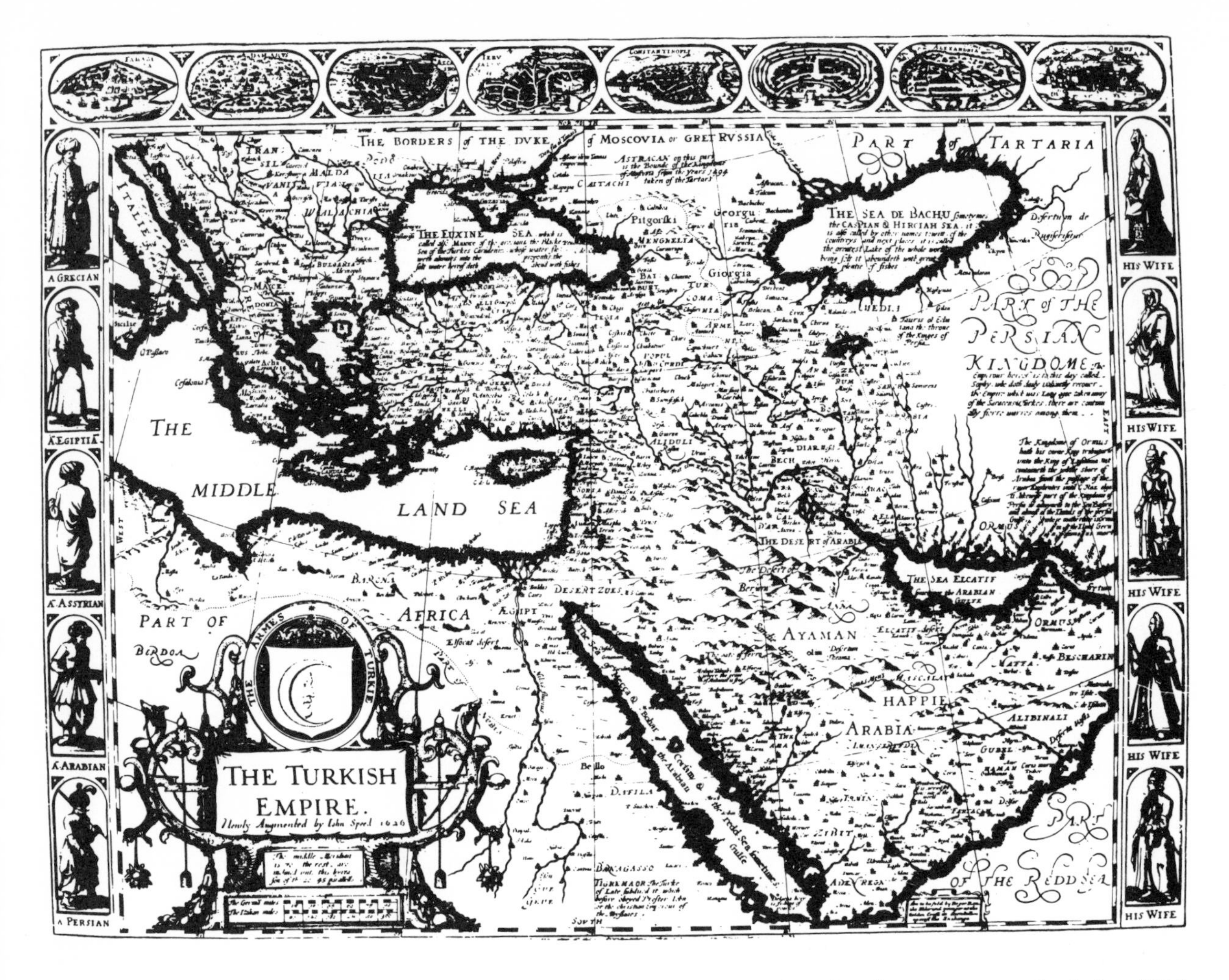

vii The Turkish Empire as recorded in 1626 by John Speed

following the pattern of the early Islamic conquests, had, by 1811, so much alarmed the Sultan of Turkey that he authorised Muhammad Ali, his Viceroy in Egypt, to mount an expedition to Arabia from the western side. An Egyptian force of ten thousand under Muhammad Ali's son Tusun Pasha, landed at Yanbo and Akaba, but was defeated in its first encounter with the Wahhabis. Not till 1818, when the formidable Ibrahim Pasha penetrated as far as Dariyah the capital, were the Wahhabis defeated, and their capital totally destroyed. Many tribes, fickle in their allegiance, were ready to support the Egyptian armies once they had arrived; such were the Bani Khalid, Mutair, Harb, Sahul and Sbei. A large measure of Ibrahim Pasha's success came from his policy of winning over even some of the converted Najdi tribes by lavish gifts and hospitality.

The destruction of Dariyah ended the first Wahhabi campaign of conquest,and Abdullah al Saud, the ruling amir, after surrendering to the Turks, was later taken to Constantinople and beheaded.

In the perspective of history the Wahhabi movement is of significance for three main reasons. Its chief importance lay in the fact that the new religious zeal was exploited by the Al Saud family to further their own ambitions. Although temporarily subject to Egypt after

1818, the family enjoyed great prestige in central Najd and were strong enough to succeed in a new bid for independence some years later. Wahhabiism had demonstrated that in central Arabia religion was still a force which could be channelled into military campaigns to further political power. Secondly, the phenomenal and alarming strength of the movement had provoked the Ottoman incursion into Najd. Before the 1811 expedition no part of Arabia except the holy places and the Red Sea coast down to Yemen was claimed by the Turks, but after the defeat of the Wahhabis Najd remained a tributary province of Egypt for twenty-three years. Finally, although support for the extreme Wahhabi creed was by no means universal, the movement left its mark right across Arabia, influencing even moderate Muslims to a more strict observance of their faith. In Kuwait as late as the 1930s, the legacy of Wahhabi ideas (which had by then received fresh inspiration) was still apparent in many aspects of daily life: the five daily prayers and the Ramadhan fast were universally and rigidly observed, women were strictly veiled and secluded, burial-grounds were simple, with no monuments to the dead, mosques had no tall minarets or domes, every man's word was his bond, the general level of honesty was exceptionally high and, a minor but significant mark of Wahhabi influence, in the speech of the bedu the name of Allah occurred in almost every phrase.

The founding of Kuwait

The founding of the original settlement of Kuwait is believed to have taken place about 1710. In the early years of the eighteenth century a group of pure-bred bedu of central Arabia were, according to tradition, driven by drought to leave their own lands and move in search of water and pasture. After much wandering, this group, who were of the Dahamshah section of the Amarat, a subtribe of the Aniza confederation, arrived at last on the southern shore of what is today Kuwait bay and, finding a good supply of sweet water, there they stayed. Among them were the ancestors of the Al Sabah, the ruling family of Kuwait, and also of the Al Khalifah who are today the ruling family of Bahrain. Several other families later to become prominent merchants of Kuwait, such as the Al Ghanim, the Shemlan and the Al Saleh (today known as the Al Mulla), also trace their origin to those early settlers.

The name Kuwait, the arabic diminutive of *kut*, a fortress, probably referred originally to a small fort built to guard the early community when they found conditions favourable and decided to make what had at first been a camping-site into a more permanent home. It has been suggested that a Portuguese fort stood on the site of Kuwait town, but of this no evidence remains, although Niebuhr states that there was one "near Graen" (the name by which Kuwait was known among early European travellers).

In the eighteenth century, the area which is today Kuwait state formed part of the territory of the Bani Khalid, the powerful Hasa tribe which dominated north-east Arabia. At the time of the arrival of the Aniza colonists there were doubtless other isolated groups settled here, probably fishermen, but in due course the authority of the Sabah, leaders of the new arrivals, was accepted by all who lived in the area round the bay. The early settlers were called Bani Utub, but the late Shaikh Abdullah al Salim has stated that this was not the designation of a clan, but merely the name by which the colonists came to be known after their long migration, meaning "the people who moved, or trekked".

Kuwait tradition has it that in about the middle of the eighteenth century the new settlement, anxious lest the Turks, their powerful neighbours in Basra, might seek to drive them away or demand tribute, sent a deputation to explain that they desired only to live peaceably. The leader of this mission was a certain Sabah, who, having shown himself competent and worthy of respect, became the first acknowledged shaikh of the community in 1756. From that time on the Sabah family have ruled continuously over Kuwait.

Little is recorded about the early years of Kuwait, but Carsten Niebuhr, who visited it in 1764, has left a valuable description of the community as he found it. He estimated its population at perhaps ten thousand, and tells us:

SHAIKHS OF KUWAIT

Sabah I	1756–1762
Abdullah I	1762–1812
Jabir I	1812–1859
Sabah II	1859–1866
Abdullah II	1866–1892
Muhammad	1892–1896
Mubarak	1896–1915
Jabir II	1915–1917
Salim	1917–1921
Ahmad	1921–1950
Abdullah III	1950–1964
Sabah III	1964

'Koueit, or Graen as it is called by the Persians and Europeans, is a seaport town, three days' journey from Zobejer or old Basra. The inhabitants live by the fishery of pearls and of fishes. They are said to employ in this species of naval industry more than eight hundred boats. In the favourable season of the year this town is left almost desolate, everybody going out either to the fishing, or upon some trading adventure. Graen is governed by a particular Shiech, of the tribe of Othema, who is a vassal of the Shiech of Lachsa (Al Hasa), but sometimes aspires at independence. In such cases, when the Shiech of Lachsa advances with his army, the citizens of Graen retreat with their effects into the little island of Feludsje."

From this account it is clear that the settlement had grown rapidly in wealth and importance if we accept the date of its foundation as barely fifty years earlier. Niebuhr's "Shiech of Lachsa" would have been the Shaikh of the Bani Khalid, who, not unnaturally, must have tried to assert his authority over the prosperous little town. However, the Sabah succeeded in establishing their independence, possibly by matrimonial alliances with the Bani Khalid.

Among the colourful semi-legendary stories of the early years of the settlement is the tale of Shaikh Abdullah I's beautiful daughter Mariam, who was wooed by a shaikh of the Bani Kaab of southern Persia. Mariam's father was unwilling to give her in marriage to the foreign shaikh, and his refusal is said to have provoked a sea attack by the Bani Kaab which was dramatically defeated by a much smaller Kuwaiti fleet, led by Mariam's cousin Salim. The traditional war-cries of Kuwait bedu *Ana akhu Mariam*, I am a brother of Mariam, and *Awlad Salim*, sons of Salim, supposedly relate to this episode in Kuwait history.

During the years of the Persian occupation of Basra, from 1775 to 1779, Kuwait enjoyed a considerable commercial boom. It had for some years been used as a transit port for goods passing from India to the Mediterranean, but now the much larger volume of trade which had previously gone through Basra was diverted through Kuwait, and many Basra merchants came to carry on their business in Kuwait. During this period, the East India Company's overland mail from the Gulf to Aleppo and Constantinople, which had till then gone from Zubair, was being dispatched from Kuwait. It was carried by camel-riders who made the journey to Aleppo in fourteen to twenty days. By contrast with this express "Desert Mail", the camel caravans laden with goods took about eighty days on the same route, or thirty days from Kuwait to Baghdad. The charge for a camel-load of 700 English pounds, covering necessary gifts to shaikhs en route, was 130 rupees. These trade caravans often consisted of as many as 5,000 camels and 1,000 men.

In about 1766, also during the reign of Abdullah I, the Al Khalifah family, kinsmen of the Sabah, had left Kuwait and moved southwards to Zubara in Qatar. In 1783, the Sabah from Kuwait assisted the Khalifah in driving the Persians from Bahrain, and the Khalifah became rulers of those islands. The Khalifah now had control of the busy maritime trade of Manama, the most important entrepot of the Gulf at the time, and it is more than likely that they allowed the Kuwaitis to have a share in the lucrative freight trade based on their port.

After Basra was recovered by the Turks from the Persians, the East India Company's resident, Mr Samuel Manesty, having some difficulty with the Ottoman authorities, moved his staff and factory to Kuwait in April 1793. The fact that they could escape the obstruction of Turkish officials by coming to Kuwait clearly demonstrates the country's independence from Turkish authority. Among those who came to Kuwait with Manesty was Brydges, the historian of the Wahhabis.

One of several Wahhabi attacks on Kuwait took place during the period when the East India Company's men were there. Manesty had some cannon brought ashore from one of the company's ships, and sepoys were stationed outside the town where the Europeans were camped. A force of five hundred tribesmen was successfully repulsed by the sepoys and the cannon, and as the attackers retreated along the shore some were killed by gunfire from the ships. In retaliation the

13. Family tree of Kawait's ruling family, starting with Shaikh Sabah II 1859–66, as drawn by a local historian

Wahhabis began to intercept the company's mail couriers, so making the route from Kuwait unsafe. The company's officials left Kuwait in August 1795 and the mail was then once more dispatched from Zubair. Until this time the Wahhabis do not seem to have interfered with the desert mail; in fact the Wahhabi amir had earlier undertaken to protect it as long as he was at peace with the Ottomans.

In 1805 Shaikh Abdullah made contact with the British, sounding them out as to whether they would guarantee him and the ruler of Zubara a safe retreat in Bahrain if they left the mainland to escape Wahhabi pressure. The implication was that the Wahhabis might otherwise force them to act against British trade. The British discouraged the Kuwaiti ruler's suggestion and rejected his overtures again in 1809 when Shaikh Abdullah offered assistance in an expedition against the Qawasim pirates of Oman. British policy was to ensure that peace was maintained in Gulf waters, while interfering as little as possible with the status quo, particularly in the relationships of the various Gulf principalities with one another.

At the beginning of the nineteenth century when pirate activities were at their peak in the Gulf, it was a Kuwaiti, Rahma ibn Jabir, who became the most notorious of those who terrorised local shipping. He prudently avoided a confrontation with the British, but was ruthless in his depredations against the ships of Kuwait, Basra, Bahrain and Muscat. With his fleet of five or six large vessels he was the most powerful and successful of the individual pirate chiefs, and pursued his career unchecked for twenty years or more, until he blew up his own ship in 1818 after a losing battle with a fleet from Bahrain.

We have little information about Kuwait for a period of forty years or so after Brydges was there on the East India Company staff. Stocqueler, a traveller who visited it about 1830, described the town as extending for a mile along the shore. He says Kuwait was governed by a shaikh who kept no armed force but levied a duty of two per cent on imports.

By this time Shaikh Jabir had succeeded his father Abdullah. During his reign he maintained friendly relations with the British, and seems to have done the same with the Turks. When the Egyptians, embroiled in warfare with the Wahhabis in 1838–9, advanced across Arabia and reached the Hasa coast, they placed an envoy in Kuwait with the obvious intention of binding Kuwait more closely to the Ottoman Empire. But while the Turks may have looked covetously at Kuwait's prosperity, there was no justification for interference while Sabah rule preserved peace and order in the state.

14. Kuwaiti Boatbuilders

In the 1860s, Col Lewis Pelly, then Political Resident in the Gulf, described Kuwait as "a clean active town, with a broad and open main bazaar, and numerous solid stone dwelling-houses, containing some 20,000 inhabitants, and attracting Arab and Persian merchants from all quarters by the equity of its rule".

By the time of the second Shaikh Sabah, 1859–66, we have the evidence of W. G. Palgrave as to the reputation Kuwait enjoyed as a seaport. Palgrave, who in 1862, made the journey from Syria to Hail, and thence south through Riyadh to Hasa, writes of Kuwait:

'Among all the seamen who ply the Persian Gulf, the mariners of Kuwait hold the first rank in daring, in skill, and in solid trustworthiness of character. Fifty years since their harbour with its little town

15. Kuwaiti dhow in full sail

was a mere nothing; now it is the most active and the most important port of the northerly Gulf, Aboo-Shahr (Bushire) hardly or even not excepted. Its chief, Eysa, enjoys a high reputation both at home and abroad, thanks to good administration and prudent policy; the import duties are low, the climate is healthy, the inhabitants friendly, and these circumstances, joined to a tolerable roadstead and a better anchorage than most in the neighbourhood, draw to Kuwait hundreds of small craft which else would enter the ports of Aboo-Shahr or Basra. . . . In its mercantile and political aspect this town forms a sea outlet, the only one for Jabal Shammar, and in this respect like Trieste for Austria, Kuwait is only fifteen days' distance or thereabouts from Hail."

Although Palgrave is inaccurate in nominating Kuwait's shaikh of the time, he appreciated Kuwait's important advantages as a port (Pelly had done the same), and must be given credit for putting on record the excellent and stable administration it enjoyed under its Sabah rulers.

The line of succession of the first five rulers of Kuwait was from father to son, but hereditary succession has never been considered obligatory. On the death of an amir, his successor was chosen by his people in the same manner that beduin tribes traditionally chose their chiefs, by selecting a member of the ruling family who had already proved himself to have the requisite qualities of personality, leadership, and good luck.

The rulers of Kuwait – in the days before oil brought sophistication to both the economy and the character of the country – had to govern two distinct groups of people whose interests were widely different, and often conflicting. The town population, except in the early days, probably always outnumbered the bedu, and from their activities the place derived its prosperity and the ruling family its revenue. But the beduin population, though poorer and fewer in numbers, enjoyed great prestige. By Arab tradition they embodied all the most admirable

16. *Main branches of Kuwait's ruling household*

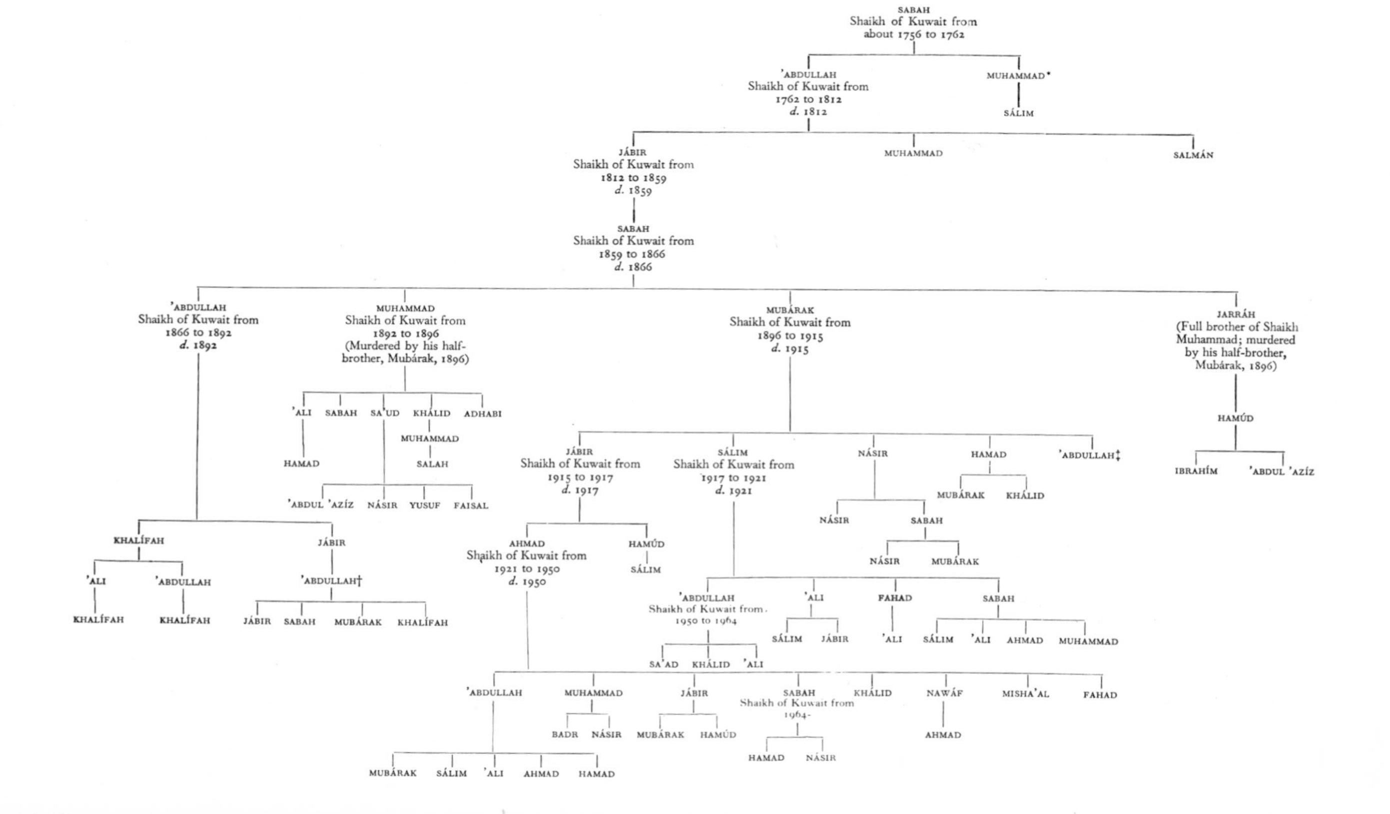

manly virtues, upholding the desert code of honour and chivalry, and constituted the fighting strength of the state. Among the settled community many prominent merchants whose families had long since given up the desert life took pride in their descent from the free-roaming tribes, though some townsmen, while they feared the bedu, would often mock them as being wild and uncouth.

The shaikhs of Kuwait, never forgetting their own beduin origins, always maintained traditional links with the tribesmen. They had to combine a beduin knowledge of the desert and a merchant's knowledge of town affairs. From their training and upbringing their outlook was more akin to that of a desert chief than that of a town merchant. Indeed, as leaders of their beduin armies in battle, they had to be completely familiar not only with the terrain of the desert, but with every aspect of tribal life. To keep the fiercely independent nomads under control it was essential for a settled shaikh to treat them with the respect due to their aristocratic standing; to sit in open majlis so that he was accessible to even the humblest of them, to be strong enough not only to exact tribute but to execute justice with speed and equity, and to be generous with gifts and hospitality to beduin leaders. Thus treated, the desert dwellers would acknowledge the shaikh's authority, and pay him the honour which they would give to a great tribal leader.

It was normal practice among the shaikhs until comparatively recently to go out to camp in the desert in spring, attended by their personal retinue of beduin guards. The Sabah, like all rulers in Arabia, owned large herds of camels, and took a natural interest in everything to do with desert life. It was customary for the amir and the male members of his family to take wives from among the women of noble tribes. Often too, a young son of the ruling family would be sent to live with the bedu, so that he might come to know the true nomadic life and share its hardships and its comradeship. In this way a boy would learn the traditional lore of the bedu and their uncompromising moral code; he would come to understand the desert man's dependence on his livestock, and the importance of the search for pasture; he would learn manly accomplishments such as horsemanship, and skill with a rifle and in the chase.

By such personal contact, and by their matrimonial alliances, each successive generation of the Sabah kept in touch with their own desert roots, and fostered the good will of the bedu – good will that was essential to maintain the order and stability on which the prosperity of the town depended. The tribal man despises a townsman who enjoys only the soft and sedentary life. The rulers of Kuwait have always shown a familiarity with the customs and rigours of the desert.

4. Mubarak the Great

17. Mubarak the Great

The Turks have already appeared in the story of Arabia as the suppressors of the first wave of Wahhabiism. Though they had not intervened in central Arabia until that time, they had for two hundred and fifty years before been active in the Gulf, where they had unsuccessfully challenged Portuguese supremacy in the sixteenth century. They had seized Basra and Qatif in 1550, and though the Portuguese subsequently expelled them from Qatif they remained at Basra, and exercised a nominal control over most of the area which is now Iraq.

From its earliest years Kuwait had mercantile associations with Basra, and relations with the neighbouring Turkish authorities appear to have remained cordial for more than a hundred years after the mission of the first Shaikh Sabah in 1756. Kuwait had permitted Turkish forces from Baghdad to march through her territory on their way to attack the Wahhabi forces in Hasa in 1798, and in 1838 the Egyptians had left an envoy in Kuwait when they, in their turn, were engaged in a later campaign in Hasa. In 1871, Kuwait's Shaikh Abdullah II gave assistance to yet another Turkish expedition to Hasa.

On the death of Shaikh Abdullah II in 1892 there followed four years when the Ottoman Empire aspired to increase its power in Kuwait. The Kuwaiti amir of the time was Shaikh Muhammad, who was weak and indecisive and allowed himself to be influenced and guided by an Iraqi, Yusuf bin Abdullah al Ibrahim, who was undoubtedly in the pay of the Turks. This Yusuf was a young man, not without worldly charm, but while he played the courtier, he was in reality plotting to bring Kuwait under Turkish subjection in the hope that the Sabah would be ousted and he himself installed as Turkish governor there. Muhammad's brother Jarrah, who, next to Yusuf, was his closest confidant, showed no more strength or competence than the Shaikh himself, and the affairs of Kuwait, particularly in the hinterland, were in a state of lawless confusion.

When political authority is weak the bedu will always assert their independence, raiding and plundering at will, and under Muhammad the Shammar and Dhafir entered Kuwait unchecked in the west, while Kuwait's own bedu, the Mutair, and the northern Ajman were terrorising more peaceful inhabitants on all sides. Indeed, these last two tribes were sufficiently defiant to practise the traditional desert "protection racket", extorting *khawah* or payment for protection, on all who travelled into or out of Kuwait town.

Turbulent times, in Arabia at any rate, tend to call forth a leader to meet the challenge of the moment, and this crisis was no exception. Shaikh Muhammad had a younger half-brother named Mubarak, who in character and outlook was the very opposite of the effete ruler. Mubarak had been reared among the bedu; he was physically tough from desert life, proud and independent in spirit, a man capable of quick decisions followed by effective action. He hated Muhammad's pro-Turkish adviser, and felt bitter about the anarchy which prevailed in Kuwait under his brother's incompetent leadership. Though a younger brother normally pays respect to an elder, Mubarak spoke out fearlessly in open council against this disastrous state of affairs. His constant criticism finally became intolerable to Muhammad and Jarrah, and to be rid of him they sent the young Mubarak out into the desert with instructions to restore order if he could – cynically failing to provide him with any funds for his allotted task. Perhaps they hoped that some marauding band of raiders would murder their troublesome kinsman.

Meanwhile, Yusuf al Ibrahim, having been instrumental in the removal of Mubarak, insinuated himself into the position of Chief Wazir of the state. Mubarak was under no illusions as to the part Yusuf had played in his removal from town, and secretly resolved to bring about the downfall of the hated Iraqi.

Out in the desert, Mubarak chafed under the impotence which came from lack of money. It was impossible to act without ammunition or supplies, not to mention payment for the few faithful followers who had accompanied him into the interior. But his knowledge of the desert and its people stood him in good stead, and his own courage and strength of personality rallied to his support some of the boldest Ajman and Rashaida leaders. Having once gathered a small nucleus of able men he successfully carried out a series of lightning raids against the more rebellious tribes, striking deep into enemy territory with great daring and skill. Though these initial successes won the admiration of the bedu, it was obvious that he could not continue without funds, and on regular visits back to Kuwait he appealed to his brothers for financial support and assistance. All his requests were denied, and to Mubarak's proud spirit it seemed there was only one course left to cure the ills of his country. Muhammad must be killed, and with him Jarrah, and Yusuf the foreign traitor.

In the spring of 1896 Mubarak began to make plans to seize power. He knew that any open attack was impossible; not only were his own men and resources inadequate, but the powerful pro-Turkish ruling clique in the town would oppose him with all available force. He must work in secret, with only a handful of trusted companions. His two sons, Jabir and Salim, scarccely grown to manhood but sharing their father's boldness and daring, were taken into Mubarak's confidence. With these two, and seven faithful tribesmen of the Ajman and Rashaida, he worked out a course of action.

The coup was to be carried out in the hot days of summer. In Kuwait town at this season Muhammad and Jarrah, like the rest of the population, would be sleeping out of doors on their roof-tops, and while they were thus sleeping, they would die. Mubarak was, of course, familiar with the lay-out of the group of Sabah houses which clustered together like a citadel on the hill in the centre of the town.

He knew it would be impossible to enter the Shaikh's residence by the one main entrance which would be closely guarded, but he also knew that from the women's quarters of a neighbouring house there was access to the rear of the Shaikh's house.

On the evening of 17 May 1896, Mubarak and his small band rode on their swiftest camels to Kuwait town, reaching the gate in the old town wall just before midnight. They were admitted by the guards, and after leaving their camels near the Suq al Madfaa, made their way on foot to the house which Muhammad and Jarrah occupied.

Unobserved, Mubarak crept through the adjacent building into his brother's house. Each member of his band knew his orders, and each proceeded to carry them out. Above all they were to make no sound, and to prevent anyone from raising the alarm. Mubarak, with one of his followers, made his way to where Muhammad lay asleep on one part of the roof, and shot him at point-blank range. The rifle shot was the signal for Jabir, who with two of the Rashaida had gone to the roof of Jarrah's apartments, to attack the Shaikh's brother with his sword. Jabir was momentarily disconcerted when he found that Jarrah's wife was sleeping by his side. She woke at the intruders' approach and threw herself full length on her husband in an attempt to save him, but Jabir's men pulled her away, smothering her screams, and Jarrah died as Jabir's sword fell. During this time, Salim and the other members of the band had made sure that none of the servants left the house or called for help.

The second objective of Mubarak's plan was to kill Yusuf, and as soon as Muhammad and Jarrah were dead, he made his way immediately to the Wazir's house. After a thorough search, he realised that Yusuf was not to be found. From the servants he learned that the Wazir had escaped by boat to Iraq the previous evening. It seemed that he had had a presentiment or warning that Mubarak was about to make a bid for power.

For the rest of that fateful night Mubarak rested and prepared to face the people of Kuwait on the morrow. As dawn approached he washed and prayed, and then went to the large majlis room of the shaikhs, which formerly stood on the site today occupied by the Seif Palace.

The story of the scene that was played out that morning in the great audience hall has been thus told by H. R. P. Dickson, who heard it from eye-witnesses:

"Dawn broke at length, and shortly after, as was the custom in Kuwait, individual members of the Shaikh's family and leading merchants proceeded in twos and threes to the great Council Hall to pay their morning salutations to the ruler. To their surprise they found none of the customary guards on duty, but instead three wild and well-armed bedu standing outside on the veranda, dangerously silent. One by one the visitors entered the majlis room, and after the usual 'Peace be upon you, O Protector', they sat down, perturbed to see only the stern-faced Mubarak seated on the ruler's throne with his son Jabir by his side. Silently man after man filed in, giving greeting but receiving none in reply and fearing to leave again. At length, when the room was full, Mubarak slowly drew his sword from its sheath and laid the bare blade across his knee, then, glancing round the assembly with eagle eye, he said in harsh quiet tones, 'O ye people of Kuwait, and blood-relatives of mine, be it known to you that Muhammad and Jarrah my brothers died last night, and I rule in their stead. If any man has anything to say, let him stand forth and say it'."

Though stunned at the news, the Shaikh's family and leading citizens accepted the *fait accompli.* Many who were tired of Muhammad's vacillation welcomed the seizure of power by a man who already had a reputation as a strong leader. The news spread rapidly from the audience hall through the streets of the town, and brought crowds flocking to the majlis, where the assembled citizens acclaimed Mubarak as their new ruler.

Once he was installed as shaikh, with the support of the townspeople and many of the bedu, Mubarak made it clear that Kuwait was an autonomous state and owed no allegiance to Turkey. But Turkey, while prepared to bide her time, had no intention of allowing Kuwait to slip out of her sphere of influence. Hoping to gain favour, in January 1897 the Turks bestowed the title of Qaimaqam on Mubarak, but he would have none of it. The following month a Turkish quarantine official arrived in Kuwait, on the tacit assumption that Turkey still enjoyed certain rights in the territory. Later, after Shaikh Mubarak had reorganised the Kuwait customs post and imposed a levy of five per cent on all imports, including those from Turkish ports, a Turkish harbour-master and five soldiers were sent from Basra to take control of the port at Kuwait, but the Shaikh would have nothing to do with them. Interpreting these gestures as threats to Kuwait's independence, Shaikh Mubarak in both 1897 and 1898 made approaches to Britain, asking for protection. But Britain at that time was not disposed to interfere between Kuwait and Turkey.

British relations with Turkey had been friendly during the preceding two hundred years, and Britain had enjoyed special trading facilities in Mesopotamia. In return, she had refrained from offending Turkey. For example, after Turkish protests, the British India steamship service which had started calling at Kuwait in the 1890s was suspended, and later Britain showed reluctance to establish a post office. But at the end of the nineteenth century Germany was actively furthering her trade in Turkish domains, and openly competing with Britain for favour and privileges from the Turks. When, in 1899, Germany obtained a concession from Turkey to build a railway from Constantinople to Baghdad, and possibly on to the Gulf, it was obvious that Kuwait, lying at the head of the Gulf, would be the natural terminus for such a railway. The dangerous political implications were not lost on Britain.

A few years earlier Russian attempts to acquire a "warm water port" in the Gulf had led Britain to seek firmer undertakings from the Arab principalities. In 1891, Muscat had signed a treaty with Britain, and in March 1892, the shaikhs of the Trucial Coast signed a similar agreement, by which they undertook (1) to conclude no other treaties except with Great Britain, (2) to admit no foreign agents except with the permission of the British government, and (3) to alienate no territory except with the permission of the British government. A week later Bahrain concluded a treaty in similar terms.

At the time of these treaties with the other Gulf states, Britain had not sought to form an alliance with Kuwait for fear of offending the Turks, but the prospect of the German railway, coming after signs in 1898 that Russia, having switched her interest from Muscat to Kuwait, was investigating the possibility of a coaling-station there, forced Britain into action. On 23 January 1899, Kuwait and Britain signed a treaty similar to those already signed by the other states in 1892. By additional agreements the Shaikh of Kuwait undertook in February 1904 not to allow post offices other than those appointed by India to operate in his territory, in July 1911 not to grant sponge or pearl concessions except with British consent, and in October 1913 not to grant oil concessions except with British consent.

Shaikh Mubarak's treaty with Britain had the predictable effect of rousing Turkish resentment, but no action followed for three years. Then in December 1901, the Turkish sloop *Zuhaf* arrived at Kuwait with an ultimatum for the Shaikh, demanding that he should receive a Turkish military detachment at Kuwait, or abdicate and retire to Constantinople. Confident of British protection if he should need it, Shaikh Mubarak replied firmly in the negative, and the *Zuhaf* withdrew without taking further action. Within the same month the British navy sent three cruisers to Kuwait, and a small detachment of troops was landed at Jahra, to forestall plans for a combined attack from the desert by Ibn Rashid (the ruler of Hail) and a Turkish force from Basra.

Yusuf al Ibrahim in Basra was still intriguing against Kuwait, and in 1902, with his encouragement, the two sons of Jarrah attempted a coup to overthrow Mubarak. Embarking at Dorah on the Shatt al Arab with a sizeable band of Sharifat Arabs, they planned to sail to Kuwait, kill the shaikh who had murdered their father, and seize power. But the British sloop HMS *Lapwing* received news of their movements on 3 September and proceeded at once to Kuwait to warn Shaikh Mubarak. Upon arrival the British ship found the town already alert and under arms. Two days later the enemy were intercepted at sea by *Lapwing* and two large baghalas containing a hundred and fifty armed men were pursued and captured after a sharp fight.

After this abortive expedition, no further attempts were made either by the Turks or by disaffected Kuwaitis to overthrow Shaikh Mubarak.

In November 1903, Lord Curzon paid a visit to Kuwait during a tour of the Gulf, and the personal meeting of the Viceroy and Shaikh Mubarak marked the consolidation of Kuwaiti–British friendship. In June 1904, the first British Political Agent was appointed to Kuwait and Col S. G. Knox arrived in August to take up his post, in spite of official protests from Turkey.

In 1907, Shaikh Mubarak agreed to lease in perpetuity to the British government a plot of land about two miles west of the town of Kuwait on the foreshore at Shuwaikh. This provided a safe anchorage, and according to H. R. P. Dickson "there is little doubt that the move was made as a counter to the then proposed Berlin–Baghdad strategic railway, which was to be continued on to Basra and Kuwait as opportunity offered. The fortifying of the area leased, and the turning of Bandar as Shuwaikh into a British naval base or coaling-station, would have effectually given HM Government control of the port of Kuwait, and brought the sea approaches to Ras al Kadhima, the site of the proposed railway terminus on the north side of Kuwait bay, under effective gunfire".

In taking up this lease Britain assured Shaikh Mubarak that it recognised the independence of Kuwait under its rulers the Sabah. It also acknowledged that the Shaikh and his heirs had sole control of the internal administration of the state, including the collection of customs dues. The British government would levy no customs charges in the leased area. Britain retained the right to relinquish the lease at any time, and in fact gave it up in 1922 when the danger from Turkey was past. In 1913, an Anglo-Turkish agreement defined the boundaries of Kuwait. With Kuwait town as the centre, the frontier was drawn as a semi-circle, with the northern edge on the estuary of the Khor Zubair, and the southern circumference touching the hill of Qurain. The offshore islands, including Bubiyan and Failaka were included in Kuwait territory. By this agreement the Shaikh of Kuwait was recognised as autonomous within this area, and as a tribal overlord, entitled to levy tribute over a larger area which extended westwards to Hafar al Batin and southwards to include As Safa and Wabra, ending at Jabal Manifah on the coast. This treaty between Britain and Turkey was never ratified, owing to the outbreak of war in 1914.

Personalities and politics

So much for Shaikh Mubarak's relations with Turkey and Great Britain. It is necessary to go back a few years to pick up the story of what was happening in central Arabia.

At the time that Shaikh Mubarak seized power, the Al Saud (who had re-established themselves in Najd during the 1840s under Faisal the Great) were suffering a new eclipse. The dissensions following Faisal's death had attracted a third Turkish intervention and in 1891, after twenty troubled years, the Al Saud were forced to surrender Riyadh to their hereditary enemies, the Ibn Rashid of Hail. In 1893, the Amir Abdul Rahman ibn Faisal al Saud had been granted permission to live in exile in Kuwait with his four sons. Abdul Aziz, the eldest, then aged eleven or twelve, later to become King Ibn Saud, grew to manhood at the court of Shaikh Mubarak, and a natural sympathy grew up between the two. Abdul Aziz was filled with admiration for the successful design which brought Mubarak to the throne, and later learned the ways of diplomacy and statesmanship from his friend and protector.

Shaikh Mubarak supported the Sauds in their aspirations to regain control of Najd. The Hail shaikhs had always been allies of the Turks, and Kuwait's western frontiers were insecure so long as Ibn Rashid held power in central Arabia. In the autumn of 1900 Mubarak, in support of the Saudi cause, led an expedition into the heart of Arabia. Early in 1901, he engaged the Rashid forces at Sarif, near Buraida. Mubarak was defeated, and his brother Hamud and nephew Sabah killed.

Later that year, when Kuwait was threatened with attack by combined Rashidi and Turkish forces, the young Abdul Aziz, now an adventurous youth of nineteen years, asked to be given command of a small company which might go south to create a diversion from the Kuwaiti forces if the attack came. Shaikh Mubarak granted his request; perhaps he had some idea of Abdul Aziz's true intentions.

Abdul Aziz had greater plans in mind than mere diversionary tactics. With speed and secrecy he led his small band of bedu to Riyadh, intent upon recapturing his father's capital. On the night of 14 January 1902, they entered Riyadh and gained access by stealth to the Ajlan Fort where they slew the Rashidi governor, taking possession of the city. Kuwaitis, who remembered the killing of Muhammad al Sabah, could not fail to mark the parallel between the coup carried out by their own Shaikh Mubarak, and the bold action of Abdul Aziz in similar circumstances. The young Ibn Saud had learned his lesson well.

Mubarak continued to give both active and moral support to Abdul Aziz ibn Saud during the following years, as Saudi power was extended once again over the whole of Najd. With the recapture of Hufuf from the Turks in 1913, Ibn Saud regained his hereditary domains.

18. Dr and Mrs C. S. G. Mylrea

Meanwhile, Kuwait flourished under Mubarak's wise and stable government. The weekly British India steamship service, suspended a few years previously, was resumed in 1901, and Kuwait's own shipping and mercantile trade expanded and prospered. Kuwait had gained the reputation of being a good place to live in, its citizens were honest and law-abiding, its climate – though hot – dry and healthy. To prominent citizens from Basra it offered a pleasant change from the damp and mosquito-ridden air of Iraq, and many of them built summer residences in Kuwait. One such summer visitor was the ruler of Muhammerah, a principality on the Persian side of the Shatt al Arab. Muhammerah's Shaikh Khazal had become Shaikh Mubarak's closest friend and boon companion. The *diwaniyah* he built himself in Kuwait in those days is now the Kuwait Museum.

It was through Shaikh Khazal that Shaikh Mubarak first came into contact with the American missionaries who were then working in Basra. Impressed by the help that Shaikh Khazal had received from the American doctor, Shaikh Mubarak invited the American Mission in 1910 to open a station in Kuwait. Dr Bennett and Mr John Van Ess came from Basra to negotiate with the Shaikh for a suitable site, and having acquired an area at the west end of the town, the mission was duly established in 1911, with Dr Bennett, Dr Paul Harrison, and Dr C. S. G. Mylrea in charge. From that time until Kuwait was able to build its own hospitals with oil royalties, the Arabian Mission of the Dutch Reformed Church in America provided a much-needed medical service in Kuwait. Dr Mylrea, who was English though serving with the American Mission, held his appointment at Kuwait for over thirty years, and became a much-loved figure in the town. He came to know Shaikh Mubarak intimately, and the Mission benefited from the ruler's interest and encouragement.

A description of life at the court of Shaikh Mubarak has been left by Barclay Raunkiaer, an enterprising young Dane who visited Kuwait in 1912. His purpose was to investigate the possibility of a later full-scale Danish expedition to Arabia, but the outbreak of the first world war, and Raunkiaer's own death in 1915, wrecked this plan. Shaikh Mubarak, while receiving him with formal politeness obviously mistrusted this lone traveller who proclaimed his British sympathies while carrying a letter of recommendation from the Turkish Wali of Basra. The strongly pro-British and anti-German shaikh was justifiably wary of strangers who might have undisclosed political motives for visiting Kuwait. It needed the intervention of the Political Agent, Captain Shakespear, to explain Raunkiaer's presence and allay the ruler's uneasiness.

At that time the property of the ruler and his family formed an imposing complex of buildings on the hill running back from the seafront in central Kuwait. The Shaikh's own private residence lay on the slope of the hill, connected to other buildings by a bridge over a narrow street at the back. The serai or government building stood on the seafront itself – part of it still stands today, a yellow brick edifice west of the new Seif Palace.

Shaikh Mubarak's daily routine followed an established pattern. Accompanied by a bodyguard of about fifty armed men, he came every morning to the serai, and in pleasant weather would sit on a veranda facing the sea where a secretary read his official correspondence to him, and the Shaikh dictated his replies. On a stool by his side there lay always his diamond-studded cigarette-case, and a pair of binoculars with which he enjoyed looking at ships or dhows in the bay. Mubarak permitted only British steamers to call at Kuwait, and had turned down the request of a German shipping firm to open an agency there.

19. Captain W. H. I. Shakespear

After dealing with his correspondence, Mubarak went to the Mahkama or administrative building in the bazaar, driving in his carriage drawn by two black horses. In front of the carriage walked a group of beduin guards, and behind, riding a white horse, came a tall negro in blue robes carrying a loaded Mauser rifle. In the Mahkama, a two-storeyed building with glazed windows on the upper floor, the ruler sat to discuss business matters and gave judgment in lawsuits.

When he had taken his midday meal Mubarak retired to his private quarters for the time-honoured siesta. During this time silence fell over the whole town as shops were shut, pack animals were still, and all the inhabitants slept. The siesta ended about 3 pm when the ring of the coffee mortar in the palace signalled that the household was astir once again, and shortly afterwards Shaikh Mubarak returned to the serai to hold audience for an hour in one of the reception rooms whose ceilings were decorated with female portraits. Apparently only favoured individuals were granted audience in the palace itself, ordinary citizens came to him in the Mahkama, where he went for his second daily session in late afternoon.

According to Raunkiaer, Mubarak's authority and control were apparent everywhere in Kuwait; he described the Shaikh as having the power and will to break all opposition, and for this reason public security prevailed.

But Shaikh Mubarak, strong, upright, decisive, as he was, could also be obstinate and intractable. Like many great men, he liked to have his own way and found it hard to tolerate opposition. Of the anecdotes which survive about the great ruler, there is one which illustrates both his stubborn pride and his humanity. It concerns Captain W. H. I. Shakespear, who in 1909 became British Political Agent in Kuwait. Shakespear was a hot-headed but courageous young man, perhaps inclined to stand too much on his dignity as the British representative in the area. The story goes that Shakespear, after he had been some little time in Kuwait, refused to visit Shaikh Mubarak, saying that the Shaikh should be the one to call on him.

20. Church of the American mission, as originally built.

The ruler naturally took exception to this, and expected the Political Agent to attend upon him at his palace. Relations between them deteriorated until there was a complete rift.

During this period, Captain Shakespear had one day sailed his boat out to the mail steamer, and having drawn alongside the ship, was highly indignant when an Arab sailing-boat came along on the outside, and its crew ran across his boat to reach the ship. In his rage he seized one of the Arabs and threw him overboard. Unfortunately the man, instead of landing in the water, was impaled on the sharp prongs of Shakespear's anchor, and died.

In genuine distress, Shakespear sought an audience with Shaikh Mubarak as soon as he came ashore. Once granted access to the ruler, he expressed his deep regret, and his willingness to pay the necessary blood-money in compensation. Shaikh Mubarak had never seen the proud Englishman so contrite, and the sight of this humility put the stern ruler into good humour once more. "My son," he said to Shakespear, with a broad smile on his face, "if killing one of my subjects has been the cause of your coming here to beg forgiveness of me, then much good has been achieved. You may kill a Kuwaiti every day if it brings us together like this!" From this time on, Shaikh Mubarak and Captain Shakespear were the best of friends.

While many British representatives have left little mark on Kuwait and their names are forgotten, Captain Shakespear's fame is still celebrated and his memory respected. This is not so much due to the possibly apocryphal events of the above story, as to the fact that he died in Arabia, fighting with the Saudi troops against Ibn Rashid at the battle of Jarrab in January 1915.

Ibn Saud had met Shakespear in Kuwait in 1910, and had been favourably impressed. When Sir Percy Cox, Chief Political Officer with British troops in Iraq, decided to send an emissary to Riyadh at the end of 1914 to enlist Ibn Saud's help for the British against the Turks, Shakespear seemed the obvious choice. While the latter was still with Ibn Saud, the Sultan's forces joined battle with Ibn Rashid, who supported the Turkish side, and Shakespear, who was an artillery officer, was killed when he assisted in manning one of the Saudi guns. He was aged thirty-six when he died. Fourteen years later, Ibn Saud, when asked to name the greatest Englishman he had known, answered without hesitation, "Captain Shakespear".

On the outbreak of war Turkey was in alliance with Germany. Shaikh Mubarak in Kuwait, true to his friendship with Britain, had promptly declared himself on the side of the allies. In return for his pledge of loyalty and for military assistance, the British government gave Shaikh Mubarak a written promise that in the event of their winning the war his five date-gardens in Iraq would be free of all taxes in perpetuity; his title to such gardens would be upheld; and he, his heirs and successors, would be maintained as shaikhs of Kuwait for all time by the British government.

Unfortunately, at the end of the British mandate in Iraq in 1921, the Iraqi government did not consider itself bound by the British promises on the subject of the Shaikh's date-gardens. The Iraqis persistently claimed tax on them, and challenged the Sabah title to the properties in the law courts.

The long-drawn-out dispute with Iraq over the Shaikh's date-gardens was just one of the matters that fell within the responsibility of the British Political Agent at Kuwait. When this appointment was first made in 1904, the intention was to have a British official on the spot to ensure that the terms of the Kuwaiti-British treaty were properly observed. In return for British protection, Kuwait allowed Britain (through the Government of India in the first place) to administer the foreign affairs of the state. This involved not only the disposal of rights and concessions to European powers, but also, more immediate and often more complex, advising on Kuwait's relations with Najd (later Saudi Arabia), and with Iraq.

The Political Agent, whom the Arabs from the beginning had called the *Gonsul*, also had normal consular responsibility for British subjects in Kuwait, including those from Indian and colonial territories. To assist him in carrying out quarantine regulations, he had an Anglo-Indian assistant surgeon, who acted as Agency Medical Officer and later ran a dispensary in the town. When an official post office was opened in 1915, this also was administered by the agency until 1921, though transferred to the Iraq postal authorities from 1929 to 1941. It was subsequently operated under the aegis of the Raj by Cable and Wireless Ltd and the British GPO, before becoming fully independent in 1958.

The men of the Indian Political Service from whom the agents in the Gulf states were recruited had to possess special gifts and special qualities of character. They had to be fluent linguists in Arabic or Persian or both; they had to have a genuine interest in the people of the Gulf states, a taste for work which took them away from the company of their compatriots, the capacity, both physical and mental, to endure the great hardships of the Gulf climate in the days when little could be done to temper summer's discomfort. The best of them lived up to an old fashioned ideal of duty which outweighed the inconveniences, and found in the opportunities of contact with the Arabs a satisfaction which compensated for the hardships.

21. Part of the mission residence established by the Dutch Reformed Church of America

Kuwait and the territory surrounding it was an area little known to Europeans, and to a Political Agent with a bent for exploration or research it offered unlimited scope to extend existing knowledge about north-east Arabia. Col Knox, the first PA in Kuwait, made a journey inland by camel as far as Hafar, the settlement in the Batin valley lying 160 miles west of Kuwait, which had not till then been visited by any European. Captain Shakespear ranks with some of the greatest Arabian explorers, having undertaken a three-month journey by camel from Kuwait to Aqaba in 1913.

Political Agents of Kuwait

1904–9	Colonel S. G. Knox.
1909–15	Captain W. H. I. Shakespear, CIE.
1915–16	Lt Colonel W. G. Gray.
1916–18	Lt Colonel R. E. A. Hamilton.
1918	Captain P. G. Loch.
1919–20	Captain D. V. McCollum.
1920–9	Major J. C. More.
1929–36	Lt Colonel H. R. P. Dickson, CIE.
1936–9	Captain G. de Gaury.
1939–41	Major A. C. Galloway.
1941 (part)	Lt Colonel H. R. P. Dickson, CIE.
1941–3	Major T. Hickinbotham.
1943–4	Mr C. J. Pelly.
1944	Mr G. N. Jackson.
1945–8	Mr M. P. O'C. Tandy.
1948–51	Mr H. G. Jakins.
1951–5	Mr C. J. Pelly, OBE.
1955–7	Mr G. W. Bell.
1957–9	Mr A. S. Halford.
1959–61	Mr J. C. Richmond (became first Ambassador to Kuwait).

Much of a Political Agent's work during more peaceful periods would today be classified as public relations. He had to personify the interest and watchful care of the British government while maintaining friendly personal contact with the ruler; he had to understand the problems of the various sections of the community; he had to be accessible to anyone who wished to see him, and in particular to entertain the chiefs of desert tribes, never sending away empty-handed those whose status entitled them to a gift. In more troubled times, as for instance during the border dispute between Kuwait and Najd and later in the Ikhwan rebellion, he relayed to Kuwait the advice or warnings of the British High Commissioner in Iraq, acted as neutral mediator between the parties in such disputes, and generally used his powers of diplomacy to pour oil on troubled waters – sometimes in a very nearly literal sense, as it was later to transpire.

22. Lt Col J. C. More

The brave and impulsive Captain Shakespear has already appeared in our story. He was the most notable of the early Political Agents in Kuwait, but there were others who left their mark in different ways. Col J. C. More, an Arabic scholar of no mean ability, served in Kuwait from 1920 to 1929, through the difficult years of Kuwait–Saudi relations. He was a man of rigid standards and strict principles. It is said that he never failed to change into evening dress each night, even when dining alone with his wife. When he called on the ruler he made it a matter of pride to be punctual to the very minute, and would stand outside Dasman Palace with his pocket-watch in his hand, timing himself to arrive at the top of the front steps on the dot of the appointed hour, with the purpose of demonstrating that when an Englishman made an arrangement he adhered to it with absolute precision.

Not all the men who came to Kuwait as Political Agents succeeded in making friends with the shaikh of the time. One of them was so disliked by the ruler that after his final departure, street cleaners were hired to sweep the road between the Agency and the Seif Palace, so that the dust that he had trodden should no longer pollute Kuwait.

5. Time of Trouble

Rise of a desert power

In the second decade of the twentieth century a new and dynamic force, a kind of Islamic crusade, appeared in Arabia. It was known as the Ikhwan – the "Brethren". Fifty years after the Wahhabi empire in Najd passed into decline with the death of Faisal the Great, the Ikhwan revived the puritanical zeal of the Wahhabis, and preached once again a return to the basic tenets of Islam. Believing themselves God's instrument for purging the land of moral laxity, they condemned as abominations tobacco, alcohol, prostitution, superstition, music and dancing, and anything that savoured of luxury or self-indulgence. As in the days of Wahhabiism, violence and cruelty were used against those who transgressed the stern ethics of the revived faith. Men seen smoking in the streets were sometimes shot, women out after dark and suspected of immorality were seized and beaten.

The outward sign of the convert to the Ikhwan was a white cloth or turban tied round the head over the *kaffiya* in place of the usual black band. Austerity in dress, as in the manner of life, was obligatory.

The Ikhwan attributed the rejection of Wahhabi purity to contact with the world outside Arabia, and sought to protect the original Arab faith and way of life from contamination by alien ideas and customs. The fierce xenophobia which stemmed from this outlook, and the primitive unformulated nationalism which underlay it, were more pronounced in this revival than in the earlier Wahhabiism, and made the Ikhwan more susceptible to use as a political tool in the hands of a nationalist leader.

From the beginning, one of the leading zealots of the movement was Faisal al Duwish, the paramount chief of the Mutair tribe. He was a true aristocrat of the desert, a man of passionate religious convictions bent on preserving all that seemed best in the life and customs of his people, a beduin fighter of immense courage and skill, and outside the Saudi royal family the greatest tragi-heroic figure in recent Arabian history. His dramatic career encompassed the highest wordly success: he was the adored hero of his tribe, the trusted friend of his king, a great military victor. Then came disillusionment, rebellion and ultimate downfall.

The movement of the Ikhwan, though obscure in its origins, seems to have sprung up in the region of Irtawiyah in Najd, the centre of the Mutair tribal *dira*. From about 1912 onwards, it had been taking hold of the bedu, among whom there was always a section ready to respond to fervent asceticism. Its adherents spread the new faith by force if necessary, threatening with attack those who would not join them, but the power of its message was such that those who embraced the faith, even reluctantly, were soon won over to fanatical devotion.

The movement gained such impetus that by 1915 Ibn Saud saw it as a threat to his own authority in Najd. With innate political shrewdness the Saudi leader knew that he must decide between two possible courses. He could either exercise his power as a temporal ruler and crush the Ikhwan, or join the movement, become its spiritual head, and exploit it to his own advantage. It was inevitable, given the prevailing opinions and conditions in Najd, that he would choose the second course. He joined the movement and became its imam, or religious leader. This title gave him authority to channel and direct the activities of the Ikhwan army, and also to collect the *zakat* or religious tax which good Muslims subscribe to be used for religious purposes.

In 1916, he issued the famous edict by which he required all the beduin tribes within his territory to join the Ikhwan, under threat of attack.

But once proclaimed imam of the movement, Ibn Saud's position was similar to that of a rider who has leaped on a runaway horse. From the beginning, the Ikhwan had been swept along by the uncontrollable pressure of their fanaticism, deferring to no authority except their own tribal leaders. In their ferocious reforming zeal they were likely to embark on attacks against Hejaz or Iraq, both courses which Ibn Saud viewed with strong disfavour. But if he could manipulate and control their militant fervour, the movement could prove to be the one force that would weld together the various hostile and dissident elements which had always existed within his boundaries. He began immediately to put into effect a policy which would ensure that he kept the upper hand.

Shaikhs of tribes which did not willingly choose to join the Ikhwan were required to attend a course of religious instruction at Riyadh, and if their loyalty remained suspect they were given houses in the capital and made to live there under the amir's surveillance. Meanwhile, trusted and fanatical exponents of the creed – Faisal al Duwish was one of these – were sent to the tribes whose leaders were detained at Riyadh, so that they might convert the rank-and-file to the true pure beliefs of the movement. In the new preaching much emphasis was laid on the role of the Imam Ibn Saud, who was held up as the father of his people, their spiritual as well as their temporal leader; personal loyalty to him was virtually an article of faith.

Nevertheless, the order of 1916 demanding universal adherence to the Ikhwan provoked much opposition, particularly in Hasa, where the Ajman had already rebelled against Ibn Saud in November 1915.

At this point Shaikh Mubarak of Kuwait re-enters the story. It will be remembered that he had always supported Ibn Saud as the rightful ruler of Najd. Therefore, when the Saudi forces found themselves besieged by the Ajman in Hufuf at the end of 1915, Shaikh Mubarak was ready to offer active assistance. He dispatched a Kuwaiti force to Hasa under the command of his second son Salim and his grandson Ahmad al Jabir. (Both men were later to become rulers of Kuwait.) The Kuwaitis were successful in raising the siege and later routed the

Ajman in battle. This tribe, one of the most powerful in Arabia, had for some years been a source of trouble to Ibn Saud. For his part, the Najdi leader had had little liking for them since they had deserted him on the battlefield of Jarrab in January of that year. From that time they had continually resisted his authority and his religious leadership. Now after their defeat in Hasa, Ibn Saud was ready to inflict severe punishment on the tribe to bring them under control.

But the Ajman appealed to Shaikh Salim for asylum in Kuwait territory, and to this Shaikh Salim agreed before returning to Kuwait with his successful army. Ibn Saud was incensed that the rebels had managed to escape his revenge, and considered Shaikh Salim guilty of deliberately acting against Saudi interests. For the first time since Ibn Saud had come to power there was now tension in his relations with Kuwait. It was a situation which might be mended only by the tact and diplomacy for which Shaikh Mubarak was famous. . . . But just when his pacifying influence was most needed, Shaikh Mubarak's help was not available. Shaikh Mubarak was dead.

Messengers bearing this news met Salim's army when it was still three marches from home. Kuwait had lost its greatest figure, a ruler whose name has gone into legend. Mubarak had brought to the leadership of the state not only the personal lustre of a strong character in the true beduin tradition, but also an acute mind which had a firm grasp of Kuwait's political situation, combined with qualities of wise and patient statesmanship. His close and long-standing personal links with Ibn Saud had been the basis of Kuwait-Saudi friendship since 1901. Now the Sultan of Najd was nursing a grievance against Kuwait which boded ill for the future.

Jabir, Mubarak's eldest son, followed his father as shaikh. During his brief rule of little more than a year he ejected the Ajman from Kuwait, whereupon they moved to Safwan with the permission of the Shaikh of Zubair. Some months later, during 1916, an uneasy truce was effected between Ibn Saud and the recalcitrant Ajman, who remained sullen and unco-operative.

Jabir died in February 1917 and was followed by Salim, the very man who in Saudi eyes was responsible for giving shelter to the Sultan's enemies, the Ajman. There followed a period when the Saudi and Kuwaiti rulers engaged in a typically beduin type of one-upmanship. During 1917 Ibn Saud induced the Awazim, a Kuwaiti tribe, to leave Kuwait, become Ikhwan, and settle in Najd. Shaikh Salim promptly brought the Ajman *en masse* back into Kuwait as his subjects. By later agreement Ibn Saud returned the Awazim to Kuwait, but most of the Ajman stayed put, unwilling to return to Saudi jurisdiction.

For the rest of 1917, Ibn Saud was occupied in taking sporadic punitive action against other refractory tribes which would not accept Ikhwanism, or would not have him as their imam. But by 1918 the situation had been stabilised. By then, most Najdi bedu, apart from the Ajman, had come round to adopting Ikhwanism under his leadership with some enthusiasm.

At this time Ibn Saud put into effect a new plan designed to control the movement. He issued a *fetwa* (a directive with religious authority), that good Ikhwan were to build themselves villages, settle on the land and become cultivators. The policy of instruction and propaganda which had been effectively and thoroughly carried out had succeeded so well that by 1920 the Sultan was getting a dutiful response to this order. It was a mark of his tremendous personal prestige that he managed to make some of the wildest nomadic bedu adopt a total change in their traditional way of life. Meanwhile, the loyal and reliable militants of the movement had become a formidable army of shock-troops, whose natural skill in desert warfare gained extra force from the zealous frenzy of their attack, and their fanatical disregard for personal safety.

Kuwait had its own problems during these years. Early in 1918, Britain imposed a naval blockade of Kuwait, believing that supplies were reaching the Turks in Damascus by the desert route from Kuwait. Shaikh Salim was informed in July that the friendship and protection of Britain was conditional upon his preventing in Kuwait any actions prejudicial to British interests. In the atmosphere of strained relations between himself and Ibn Saud, Shaikh Salim believed that the Najdi ruler had directly caused the imposition of the blockade by accusing him to the British of pro-Turkish sympathies. The situation caused bitter resentment in Kuwait.

Although the blockade was lifted after the armistice with the Turks, Shaikh Salim anticipated that before long the animosity of Ibn Saud would take a more direct form. He greatly feared the power of Najd now that the ruler could command the support of the Ikhwan army under its chief Faisal al Duwish, and believed the Saudi leader had designs on Kuwait territory. He fully expected that Ibn Saud would urge Britain to change her views of Kuwait's boundaries, as defined by the unratified Anglo-Turkish agreement of 1913. In that treaty, Kuwait's indirect control – with the power to levy tribute from the bedu – was recognised as extending southwards to Jabal Manifah, the two Jariyas, and Wabra, halfway between Kuwait and Qatif.

Anxiety about his southern boundary led to an announcement by Shaikh Salim that he intended building a fort near Jabal Manifah to mark the limits of his territory. Ibn Saud immediately claimed that the location in question was on his land, and retaliated by ordering an Ikhwan settlement to be built at Jariya Ilya, which was claimed by Shaikh Salim.

The Kuwaiti ruler, who was a stern and unyielding man, little given to compromise or reconciliation, in April 1920 sent his beduin troops under the war-standard of Kuwait to encamp at Hamdh fifteen miles east of Jariya Ilya, probably intending only to make a show of force.

But against the Ikhwan mere gestures carried little weight. While the Kuwait army was camped at Hamdh, Faisal al Duwish, the fierce warrior leader of the Ikhwan, made a sudden brutally effective dawn attack, almost wiping out the Kuwaitis, and escaping with many of their camels.

When the remnants of this army returned to Kuwait in the summer of 1920, Shaikh Salim at once ordered the building of a defensive wall around Kuwait town. Despite the fierce heat of summer it was completed within two months. There had been an older wall which had fallen into disrepair, but the wall built in 1920 existed until 1957, and its gates are still standing, preserved as monuments in modern Kuwait.

While Shaikh Salim and Ibn Saud both tried to justify to H.M. Government actions which had led to the battle of the Hamdh, Britain reaffirmed that it recognised the boundary laid down in the 1913 agreement. After further communications and recriminations between the Najd and Kuwait rulers, the British government agreed to appoint an arbitrator on the boundary question. At this time Ibn Saud was claiming that Kuwait jurisdiction extended no further than the area enclosed within the new town wall.

Battle of Jahra

In early September 1920, intelligence from the desert informed Shaikh Salim that Faisal al Duwish was moving north to the wells of Subaihiyah. In Kuwait it was feared that the Ikhwan, on Ibn Saud's orders, were planning an attack on Jahra, the strategically situated village which commanded Kuwait's route to Basra. The British Political Agent could not believe that Ibn Saud had ordered this attack, and was convinced the Ikhwan were acting independently and out of control. Shaikh Salim took command of all available fighting men, and proceeded to Jahra where he prepared to defend the small oasis.

The Ikhwan had developed and perfected new battle tactics, more systematic and organised than in the older form of beduin warfare. An attack such as the one on Jahra would be carried out on foot, with the men drawn up in a series of lines, in imitation of conventional military procedure. Horse and camelmen were kept in reserve. In this arrangement, the first three lines were formed by beduin "irregulars", while supports and reserves consisted of Mutadaiyanin (those of the faith), the fanatical religious elite. These latter delivered the main stroke after the first wild attack has been made by the front ranks.

The beduin goes into battle with his kaffiya muffled round his face leaving only the eyes visible. Nobody wants to be identified when it comes to killing. As the troops charge, every man takes up his tribal war-cry, or the war-cry of his overlord.

Faisal's attack on Jahra was launched on the morning of 10 October 1920. Shaikh Salim was lined up with his forces south-west of the village, with Ibn Tuwala's mounted Shammar on his right and Duaij al Fadhil's horsemen on his left. They were considerably outnumbered by the Ikhwan. Faisal's charge was directed at the western end of the line, and when the Shammar broke, he outflanked the Kuwaitis from the north-west. Fierce fighting continued for about three hours, but by then the Ikhwan were in control of the main part of the village, and Shaikh Salim, with other shaikhs and about six hundred men had been forced to withdraw into the Red Fort on the south-west edge of the palm-groves. The fort, which enclosed a rectangular area about eighty yards square, consisted of thick fifteen-foot high walls, with a round tower at each corner and one over the gateway which faced north-east. On the inner side of the wall there was a firing platform with loopholes, but the holes were badly constructed, giving only a limited field of fire.

During the afternoon a messenger from Faisal came to offer peace terms to Shaikh Salim. The Ikhwan were prepared to give the Kuwaitis safe conduct out of Jahra if they evacuated the fort, but Faisal and his men would keep all camels and booty captured in the battle. Shaikh Salim knew that the safety of Kuwait depended on his keeping the Ikhwan engaged at Jahra. He refused the offer of terms.

In early October the heat was still intense; there was no well within the fort, and the occupants were already suffering from thirst. They knew that they could not withstand a siege for any length of time, but they counted on inflicting enough damage on the Ikhwan forces to render them too weak to attack Kuwait. The Ikhwan army had taken refuge in the gardens to the north of the fort, which were enclosed in high walls giving them excellent cover against the fire of the besieged Kuwaitis. Having rejected the enemy's terms, the holders of the fort awaited nightfall, and prepared for an Ikhwan assault.

Three times during the night of the 10th, the Ikhwan hurled themselves at the walls of the Red Fort with fanatical courage; each time they were driven off by the rifle-fire of Shaikh Salim and his men. When dawn broke, the open ground outside the fort was strewn with dead or dying; the strength of the Ikhwan was broken but Shaikh Salim's casualties were comparatively few.

The gunfire at Jahra had been heard through the night in Kuwait town, where the population was in a state of great anxiety, fearing for the fortunes of its army under Shaikh Salim. Everyone knew that the Ikhwan would turn to attack Kuwait if they were not checked at Jahra. All able-bodied men left in the town rallied to defend the wall, keeping vigil on the firing-platform through that night of fear. On the morning of the 11th, Shaikh Ahmad al Jabir, who was in command in Kuwait, organised a relief expedition to go to Jahra. About six hundred men were sent by sea in the Shaikh's steam launch and several sailing boats, while another mounted party set out by road.

23. Shaikh Salim al Mubarak, the victor of Jahra

But the brave defence of the Red Fort had decimated Duwish's army. An eye-witness counted eight hundred Ikhwan dead on the morning after the assaults, more than that number were severely wounded; of these, five hundred or so were to die within the next few days. Kuwait had lost about two hundred men.

While the relief force from Kuwait was on its way, Faisal al Duwish was negotiating new truce proposals in Jahra but no longer speaking from a position of strength. The profoundly religious and moral character of the Ikhwan army is shown in the terms they now put forward, terms which would seem totally absurd in the context of most military campaigns. An *alim* or religious teacher – one Ibn Sulaiman – was sent to parley with Shaikh Salim. He called on the Shaikh to suppress smoking, drinking, gambling and prostitution in Kuwait. Shaikh Salim answered that he, like all good Muslims, disapproved of these activities and was prepared to prohibit them, though he was unable to control what was done privately in men's homes. But there was still the question of the captured camels and loot. Shaikh Salim now demanded that all the spoils must be given up, and if this was done, he would let the Ikhwan withdraw from Jahra without interference. Ibn Sulaiman was unable to agree without further reference to Duwish, and returned to his own camp to consult his leader.

Shortly afterwards, the Ikhwan army moved out of Jahra, taking all the spoils of war with them. They marched for three hours that day, and on the following day reached Subaihiyah; on their retreat they buried hundreds of their wounded who had not survived.

At Jahra, the corpses on the battlefield had to be cleared away quickly in the great heat. Most of them were disposed of in nearby wells which provided ready-made burial pits.

For some weeks after the battle, the village of Jahra was entirely evacuated and when it was reoccupied a wall was built around it to make it less vulnerable to future attack. The Qasr al Ahmar – the Red Fort – has been preserved as one of modern Kuwait's links with the past, a reminder to the new generation of the tenacity of their grandfathers in standing firm against the Ikhwan fanatics.

While the Ikhwan were at Subaihiyah, messengers came and went between Duwish and Shaikh Salim, who had meanwhile asked for British assistance. When Salim received a deputation from Duwish on 24 October, Major More, the Political Agent, was present. He repeated to the envoys the contents of a communiqué which had already been dropped by the RAF over the Ikhwan camp. The gist of this was that Britain could not stand by while Kuwait town was threatened. Any attack on Kuwait would be considered an act hostile to Britain, and would be opposed by British forces. The RAF had already been making their presence known, and British warships were anchored in Kuwait Bay. Despite all the evidence and the explicit statement of the deputation's leader that Duwish's forces were operating under Ibn Saud's direct orders, Major More persisted in believing the Ikhwan attack on Kuwait had been carried out without the Sultan's approval. The Saudi amir, who was being subsidised by Britain at the rate of Rs75,000 a month, continued to protest his innocence, while using Duwish to further his designs against Kuwait. British officials of the time, dazzled by Ibn Saud's military successes, and his charismatic personality, seem to have naively accepted his denial of responsibility.

Shaikh Salim refused the envoys' request to be allowed to buy provisions and dismissed them. On 26 October, the Ikhwan moved south to As Safa, to await ammunition and supplies.

Sir Percy Cox, the High Commissioner in Iraq, ordered that Subaihiyah wells should become a "no man's land" between the Saudi and Kuwaiti sides, to avoid their confrontation and possible further fighting before the dispute could be settled.

An uneasy truce existed through the months that followed. In February 1921, the Shaikh of Muhammerah's son Chassib, acting as mediator, went with Shaikh Ahmad al Jabir to meet Ibn Saud at Khafs north of Riyadh on 2 March. But on 4 March, news of Shaikh Salim's death reached them. He had died on 27 February.

Ibn Saud at once declared that he no longer had any quarrel with Kuwait, and that there was no need for a formal boundary between the two territories. Shaikh Ahmad al Jabir, on returning from his peace-making mission, was elected to power by the Sabah family council.

Conference of Uqair

Despite the Saudi assurance, the need to fix firm and internationally recognised frontiers between Kuwait and its powerful neighbours became increasingly obvious, and in 1922 Sir Percy Cox determined to settle the matter. Britain, it must be remembered, was subsidising the exchequer of the Najdi ruler at this time and had undertaken the role of kingmaker in Iraq. The High Commissioner at Baghdad was in a powerful position, and he was not a man to underestimate the strength of his armoury. A meeting was fixed between the interested parties for November 1922. Uqair was the venue. Ibn Saud informed the British that he would arrive in Uqair on the 21st of that month.

Thus, final plans were made for a crucial desert meeting.

Sir Percy went via Bahrain aboard one of HM sloops, accompanied by the Iraq delegate Sabih Beg; Major J. C. More, the Political Agent in Kuwait representing Shaikh Ahmad; Shaikh Fahad Beg al Hadhal, head of the Amarat, and various assistants. There he was to be met by Col Dickson and the party would cross to the mainland meeting place. Ibn Saud's party had already arrived when the British–Iraq contingent reached Uqair. It included his brother-in-law and one-time adversary Saud al Arafa al Saud; a prominent Najdi merchant, Abdal Latif Pasha al Mendil; the Lebanese poet and historian Amin Rihani, and several other officials, as well as a personal bodyguard of some three hundred men.

The Uqair Conference was of paramount importance to Kuwait and, indeed, to the whole of the Arabian peninsula.

Iraq, with King Faisal newly on the throne, wanted to establish its boundaries and thus gain recognition by the League of Nations. Sabih Beg was clear as to where the rightful frontiers lay. He indicated an area which ran to within twelve miles of the Saudi capital Riyadh, thus taking in nearly the whole of the northern half of the kingdom of Arabia. He also designated a western boundary on the Red Sea and the Gulf region extending to Qatif. "As God is my witness, this and only this is the true boundary and cannot be disputed," said Sabih Beg.

Ibn Saud angrily challenged the geography and the historical validity of Iraq's claim. "From the days of Abraham, my ancestor, the territories of Najd and the bedu world have extended as far north as Aleppo and the river Orontes in northern Syria, and include the whole country on the right bank of the Euphrates, and from there southwards to Basra and the Persian Gulf." Iraq's claim to a large slice of Arabia was thus countered by a demand for most of Syria, a large part of Iraq and all of Kuwait. The Kuwait Political Agent is said to have maintained utter silence throughout the proceedings. Sir Percy Cox, in Col Dickson's presence, took aside Ibn Saud and said that he would have no more of "these impossible arguments and ridiculous claims". He, Cox, would decide the frontiers. The High Commissioner whose powers of persuasion won the admiration of all who saw and heard him during his working life in Persia, Arabia and Iraq, had the last word. He called the delegates together and traced a line on the map from the northern tip of the Gulf to Transjordan. It cut off some of the territory demanded by Ibn Saud, though some 300 miles less than Iraq was claiming. Sir Percy reprimanded the Sultan of Najd as though he were a naughty schoolboy. "It was astonishing," remarked Dickson.

24. Major-General Sir Percy Cox, Britain's High Commissioner extraordinary

The High Commissioner also drew a line round two arbitrary areas at the southern and western extremities of Kuwait. These were to be called the Kuwait and Iraq Neutral Zones, providing a buffer between all the adjoining states and giving them free and equal access to fresh water and grazing within their boundaries.

The new borders of Kuwait were defined by a line which started at the junction of Wadi al 'Aujah and the Batin valley in the west to the point where the 29th parallel of latitude met the border of Najd and Kuwait as shown in the Anglo-Turkish "red line" oil agreement. It then followed the same red line to a coastal point just south of Ras al Qalai'ah (Jilai'ah). South of this boundary was the Kuwait Neutral Zone.

"To placate Ibn Saud," wrote Dickson, "Sir Percy deprived Kuwait of nearly two-thirds of her territory and gave it to Najd."

There was an interesting postscript to the Uqair Conference. Abdul Latif Pasha al Mendil asked Sir Percy Cox why he thought the neutral zone necessary. Sir Percy replied that the Kuwait tribes must have adequate grazing. Then he asked, "Why, pray, are you so anxious that this area go to Najd?"

"Because we think there is oil there," replied the Pasha.

"That is exactly why I have made it neutral," said Sir Percy. "Each side shall have a half share."

Frontiers may have been the ostensible subject of discussion, but oil was never far from the thoughts of the participants. Indeed, that subject was highlighted by the presence of an interloper who thought of little else, Major Frank Holmes, a short, stocky, awkward and amusing man whom Dickson and, indeed, the leaders of Saudi Arabia and the Gulf States, were to see a great deal of in the years ahead. With characteristic aplomb he turned up at this important political conference unannounced and stationed himself between the two camps – though a little closer to the Arabs than the British, it was said. He had talked to the Najdi leader prior to the Uqair meeting and Cox was suspicious.

The conference disbanded, Sir Percy Cox told both Ibn Saud and Holmes to be circumspect in any oil talks they might have. Nothing could be settled, he said, until the British government had been consulted. The Sultan of Najd expressed his regret to Holmes and the parties went their various ways.

Sir Percy, dressed as ever in diplomatic suit and black homberg, went on to Kuwait to tell Shaikh Ahmad the details of the agreement. The ruler, as might have been expected, received the news with dismay. Several years later the Shaikh told Colonel Dickson that the Uqair decisions had shaken his faith in Britain. It was, he said, the sacrifice of a small nation to a greater power. The kind of thing he believed Britain to have gone to war to prevent.

A troublesome period was in store for Shaikh Ahmad, who had been in power for no more than a year when Kuwait's boundaries were constricted.

The long blockade

It has been necessary to interrupt the story of Kuwait's history to follow the events at Uqair, where vital decisions on the state's boundaries were being made. But let us return to Kuwait itself.

Shaikh Ahmad al Jabir, who had succeeded his uncle Salim on 27 February 1921, was a good-looking genial man of considerable personal charm. Strictly religious like all his countrymen, he devoted himself sincerely to furthering the best interests of his state and people. He was well disposed towards the British, and ready to co-operate with them, but his characteristic Arab pride made it hard for him to forgive a discourtesy, and he could be awkward to deal with unless treated with due tact and diplomacy. In his long struggle to keep the state solvent through great financial difficulties from 1923 to 1937, he got the name – especially among the bedu – of being miserly. Dickson defends him against this charge, as against criticisms that he showed little real strength of will, and describes him as "a man of quietly firm character, capable of lightning decisions, whose ambition in life was to follow in the footsteps of his famous grandfather, Mubarak".

Though he was conservative in desiring to preserve traditional Arab values and customs, which he considered best for his people, he was not personally averse to western gadgets and material innovations. With the help and advice of the Political Agent and the Calverleys of the American Mission, he furnished a drawing-room and dining-room in Dasman Palace with European-style furniture brought from India. He also took a keen interest in modern guns and rifles, cameras, movie projectors and films, and cars. He enjoyed driving his own car.

At the end of Shaikh Salim's reign there had been growing dissatisfaction among the trading community in Kuwait at the continual warfare from which they suffered, and the generally unstable conditions which had adversely affected commerce.

Shaikhs of Kuwait had traditionally ruled autocratically, without reference to any form of council. In addition, the ruler used to settle personally all forms of dispute, making himself available daily to any of his subjects who had some matter to lay before him. But when Shaikh Ahmad came to power some of the merchants were expressing the view that the people would not tolerate another absolute ruler. Led by Hamad bin Abdullah al Saqar, they insisted that the new shaikh should accept a council of advisers drawn from prominent merchants, with whom he was to consult in all matters external which affected the town, being guided by their opinions in making decisions.

25. Shaikh Ahmad al Jabir

A council of twelve members, six from the east of the town and six from the west, was duly elected, with Hamad al Saqar as president. But it seems that once matters had gone this far the urge for reformation lost its impetus. In practice the council rarely met, and Shaikh Ahmad continued to rule according to the time-honoured system which the majority of the people understood and accepted.

But although this council never really functioned, it was a significant pointer to a certain restlessness among Kuwaiti merchants. Hard-headed, worldly, looking for new business opportunities, they were eager for progress and innovation which would create fresh markets in Kuwait. The Shaikh, by contrast, felt that he must preserve the Islamic character and way of life of his state, and was reluctant to see changes which might undermine the probity and morality which had always been such admirable qualities of the Kuwaitis in public and private life.

Before his accession, Shaikh Ahmad had already made a favourable impression on King Ibn Saud, and the King had welcomed the news of his nomination as shaikh. Nevertheless, after the treaty of Uqair, the King put into effect a boycott on trade with Kuwait, a short-sighted move that caused much harm to Kuwait–Saudi relations, and fresh problems to Ibn Saud himself. He banned his tribes from buying supplies in Kuwait, ordering them to use instead Uqair, Qatif, and Jubail. His action was prompted partly by a desire to keep his bedu away from Kuwaiti influences (remembering that the Mutair had formerly been subjects of the Shaikh of Kuwait), but even more by a desire to increase the prosperity of his own ports at the expense of Kuwait. Behind these considerations lay his resentment at Britain's continued recognition and support of Kuwait, which he now believed was an integral part of Najd.

H. St J. B. Philby writing from Riyadh in an unpublished letter to the Dicksons was later to comment: "Of course the whole trouble about Kuwait is that it is racially and geographically a part of this country, though it is artificially separated from it by a political barrier which the British in their folly prefer to keep up. You might just as well make Hull and its district an independent principality under German protection – it would die, as all traffic would be diverted to Harwich (Jubail) or Dover (Ras Tanura)."

This boycott interfered with the traditional patterns of migration of the Najd tribes who were in the habit of trading in Kuwait at certain times of the year. The inconvenience and hardship caused by the King's prohibition roused bitter feelings among the bedu, and it was an important cause of the build-up of dissatisfaction which was to culminate five years later in the Ikhwan rebellion.

For Kuwait the Saudi blockade was a disaster, not only because it cut off so many regular customers from Kuwait's shop-keepers, but even more because it stopped trade with the Qasim, for which Kuwait had always been the natural sea outlet. Ever since Kuwait's establishment as a port, regular camel caravans had carried much of its volume of imports to Anaiza and Buraida. Not a few of the Kuwaiti merchants had partners in these two Qasim towns, and by reason of their own Najdi ancestry and outlook Kuwait's trading connections lay in that direction rather than towards Iraq.

By 1930, a state of acute economic depression prevailed in Kuwait. After seven years of the Saudi boycott, the pearl trade, which had been the town's main industry, suffered a severe slump due to the general world economic situation, and these combined misfortunes reduced Kuwait's commercial life to near stagnation. British official records show that Kuwait's imports of tea and sugar, two of the staple commodities of trade with the bedu which represented 25 per cent of all imports before Saudi restrictions, had by 1929 dropped to less than a quarter of the volume for 1922, before the blockade began.

During those years, there were always cases of individual Saudi bedu attempting to buy supplies in Kuwait, but it was hard to escape the vigilance of Ibn Saud's guards and spies, and the savage punishment inflicted on those who were caught – summary death by throat cutting – was enough to deter all but the most desperate or foolhardy.

An interesting sidelight on the years of depression is provided by official records of 1930. During that year an agent from Muhammerah arrived to collect substantial debts owed to Shaikh Khazal by Kuwaiti merchants. Many of the merchants were unable to meet their debts and offered to surrender house property in lieu of cash. As Shaikh Khazal already owned extensive properties in Kuwait town neither Shaikh Ahmad nor the British government welcomed the prospect of his acquiring further large portions of real estate, and the Political Agent was instructed to advise against the deal. Later, the Shaikh was prepared to agree provided such property reverted to him on Shaikh Khazal's death, but the British government still opposed the idea. It is significant that in the British view the merchants were regarded as totally honest and trustworthy, and an official directive to the Political Agent expressed the opinion that the debts could well be left outstanding since the Kuwaitis could be relied on to pay up as soon as they had the available cash.

To understand the reluctance shown towards settling the debts by the transfer of property, it must be remembered that Shaikh Khazal, though he had formerly been the independent ruler of a small state at the mouth of the Shatt al Arab, was in Persian government eyes a Persian subject. If too large a part of Kuwait town came into the ownership of a Persian citizen, it was feared that this might substantiate Persian claims to sovereign rights over Kuwait. Throughout the late twenties and thirties, official documents reveal a general nervousness lest Persia or Iraq should find opportunities or excuses to interfere in Kuwait affairs, and British officials were constantly alert to forestall or prevent anything that might provoke such interference.

Old Kuwait

Though times were hard old Kuwait was, by the prevailing standards of Arab townships, a place of much activity and interest. It had a population of about 60,000 by 1930, with the pleasantly disorganised look of a community which has grown up at random. Few streets ran straight for any distance, houses were not aligned, and the sandy roadways had mostly come into existence where convenient access paths were needed for the donkeys, horses or camels which passed constantly to and fro carrying goatskins of fresh water, dried palm-fronds for firewood, charcoal, or other merchandise.

26. Kuwaiti, with hawk

Since the common building material was mud, used either as sun-baked bricks or applied wet to build up a wall course by course without pre-shaping, the general colour of the town was the same tawny ochre as the ground from which it rose. Apart from a rare tamarisk tree, it was unrelieved by any visible touch of greenery or vegetation.

The houses, built round courtyards, were mostly of one storey only, but here and there a single upper room would rise above the general level of roof-tops. Behind high windowless walls whose blank exterior was broken only by projecting wooden water-spouts, the life of each family went on in the enclosed courtyard. In an affluent household there would be not only the mistress of the house and her children but serving-girls and perhaps unmarried female relations, so that a harem was a place of sociable companionship, not, as Europeans have sometimes thought, a lonely prison. Well-to-do houses had a second courtyard so the master could entertain his friends in his own half of the house. At the entrance to such masculine apartments there was always a coffee room, and benches were often set outside the gate where the men might sit out to catch the evening breeze.

The town stretched four miles along the seashore and perhaps two miles inland at its densest central part. The poorer quarters for the most part lay at the back or landward side of the town. On the south-eastern fringe were many walled *hautas* or gardens which were used by men of prosperous families as their places of recreation, amid the scant greenery of a few *sidr* trees.

Though the fortified wall enclosed the town on the southern side, there was a wide open space between the limits of the built-up area and the wall. In this space at the east end of the town there were small market-gardens growing radishes, tomatoes and cucumbers. In other parts the ground was pitted with holes where clay had been dug for building, or where gypsum had been uncovered and burned to be excavated for plaster. Two Arab cemeteries formed wide desolate areas where graves were marked by lumps of rock on the uneven ground. There was also a small Christian cemetery in which, among a few graves, stood a memorial to Captain Shakespear. In 1952, Dr C. S. G. Mylrea was also buried there, but since that time Christian burials have been taken to a new cemetery to the west of the town.

At the western end, the Naif and Jahra gates in the town wall provided access for traffic from the desert. It was on this side of Kuwait that townspeople and tent-dwellers were most obviously juxtaposed, for there was a large encampment of black tents immediately outside the wall, and some tents even found space to pitch within the precincts of the town.

Immediately inside the Naif gate stood the rectangular walled fortress with corner turrets, the Qasr al Naif. This was the headquarters of the Shaikh's men-at-arms, and contained their arsenal of rifles. It was still standing in 1970, hedged about by tall buildings, surviving as one of the city's many administrative offices. In the centre of the courtyard is a gallows where executions have taken place in the past.

Going into town from the Naif gate and past the fort, one came to the *safat*, the spacious market square. Here the desert man did business with the town, camels and sheep were offered for sale, and caravans were loaded with goods for the interior. To the north of the square clustered the shops which catered specially to beduin needs, and among the throng there were always men from the deep desert, wild-eyed, ill at ease in town surroundings, but proud and upright in bearing, walking with the light tread of feet that have never known sandals or shoes. All of them wore the brown woollen cloak which, however ragged it might be, was indispensable to a beduin's self-respect; most of them held the thin straight cane which was part of every man's equipment in the desert; one or two of them might carry a falcon on a gloved wrist, for the skilled falconer liked to maintain personal contact with his bird.

Women too would mingle among the crowds in the streets, their figures entirely draped in black, one hand always holding the edges of their cloaks together across their veiled faces to conceal even the outline of their features. In one covered alley off the *safat* was their own market, where women shopkeepers sold clothes and trinkets. At one end of this market one might see the silversmiths working on beduin jewellery, melting Maria Theresa dollars over glowing charcoal to fashion bangles and anklets, or rings and nose-rings set with turquoise.

Kuwait's principal business quarter was concentrated in the area of the main *suq* whose covered thoroughfare ran directly through the town centre between the *safat* and the customs wharf on the seafront. In this street and the honeycomb of alleyways which ran off it, were to be found the workshops of the skilled artisans, as well as the premises of merchants selling carpets, cloth, pearls, gold, spices and cheap enamel and glass-ware. Nearer the seafront important wholesalers of rice, coffee, tea, and sugar had their warehouses. The whole market quarter assailed the senses with colourful sounds, sights and smells.

Another, quite separate, area of activity in Kuwait was the seafront. Here the road between the houses and the shore provided one of the most picturesque scenes of the old town. Along nearly the whole length of the front the sloping beach was lined with boats, among them vessels in all stages of construction. By the roadside stood steel capstans and wooden water-tanks, and everywhere there were masts and spars and piles of timber. Where the road was wide enough, the white sails would be spread out on the ground for the sail-makers to sew. On every side one could see the industry and craftsmanship and sheer hard labour which had gone into making and maintaining Kuwait as the finest boat-building port of the Gulf. The boats which took shape here were marvels of the carpenter's art, made with primitive traditional tools, but shaped into lines so functionally graceful that they were a delight to the eye.

To provide shelter for the sailing craft a series of small harbours existed all along the front, and there was a continual traffic of boats coming and going. The seafaring activity based on Kuwait in those days fell into three categories. There were the smaller boats used in local traffic, including the fleet of medium-sized *boums* which carried

27. Harbour scene in the 1930s

Kuwait's fresh water supply from the Shatt al Arab, the lighters which brought ashore the cargo from the British India steamers anchored in the bay, and the fishing fleet.

Then there were the ocean-going *boums*, the special pride of the Kuwaiti ship-builders. Some of these were as big as three hundred tons, and they overshadowed all other boats in the harbour. They were used for the annual trading voyage or *sifr* which lasted six to eight months during the winter season of the north-east monsoon in the Indian Ocean. Leaving Kuwait in the early autumn and collecting a cargo of dates from Iraq they would follow one of two traditional trade routes. The first took them down the Gulf along the coast of Baluchistan to Karachi, and then southwards to Cochin and Calicut, where they would load mostly timber for the return trip. On the second route they kept close to the Arabian coast, selling their dates in ports along the way as far as Aden. There they would take on salt, and smaller quantities of other goods such as cloth, incense, and ghee, and sail down the east African coast, often as far as Zanzibar. The principal cargo brought from there was mangrove poles used for roofing houses in Kuwait.

The sailors on the *sifr boums* were among the most expert in the east, growing up to the trade as fathers took their sons with them on their voyages from the age of ten or eleven. They were familiar with every current and shoal in the Gulf, and had an intuitive understanding of wind and weather. In the Indian Ocean they would navigate by the stars when out of sight of land, though most of them carried a compass and binnacle. Returning from Africa they sailed almost against the wind, close-hauled and demonstrating the skilled seamanship for which they were justly famed.

The owners and captains of these boats often came from old-established merchant families of Kuwait, such as the Al Ghanim, the Shahin, the Qitami, the Saqar. In many such families the household was left without a man for six months of every year, and there was always great rejoicing in the spring when husbands and kinsmen were welcomed home at the end of the *sifr*.

When the great *boums* were laid up in the harbour for the summer, held upright on their keels by supporting poles lashed to their sides, their masts were dismantled and a protective awning of matting was built over them to shade the decks from the summer sun. The hulls were cleaned, and the timbers above the water-line were treated with shark-oil until they shone a rich golden brown.

The third type of seafaring activity was pearl-diving. Before the first world war Kuwait worked as many as seven hundred boats, which would have required crews and divers totalling from 10,000 to 15,000 men. Since Kuwait could not provide this labour force an annual influx of bedu from Iraq and Najd came to chance their luck in the pearl dive. Up to the late 1920s pearling was Kuwait's most important trade, but by the end of the decade only about 330 boats were active since the slump in the world's pearl markets badly affected the industry. No exact figures exist for the value of the trade from Kuwait, but in Bahrain, the acknowledged centre of the industry, pearl exports for 1930 were worth between 1½ and 2 million pounds sterling. One may assume that for Kuwait in the same year the figure would be somewhere near a million pounds.

The pearl-diver had to be an adventurer, since divers drew no wages, only getting a share in the profits if the boat made a good catch. Needless to say, many were unfortunate, and if their boats had a bad year there was no reward for their toil, and they were left with a debt to their captain who had supplied their provisions, a debt which bound them to serve with the same captain in the following year.

The Kuwaiti boats traditionally worked in an area about a hundred miles south of Kuwait, where the depth of water was around forty feet. There was a strong sense of fraternity in the whole pearling fleet and boats stayed close together throughout the season. This lasted from roughly mid-May to mid-September, the hottest time of the year, so that life for men on the boats which lay in a dead calm sea under a blazing sun was extremely harsh. The pearling boats varied from thirty to sixty feet in length. They were always crowded with men, the larger ones carrying as many as two hundred.

The divers, for fear of fatal cramps, lived on a minimum of food for the whole season. Sickness, skin troubles and scurvy were com-

28. Fish trap, an important source of food during the long blockade

mon, and they also knew the dangers of diving too deep and surfacing too rapidly. Many died young, their health irretrievably ruined by malnutrition and the rigours of pearling.

The manner of diving as practised in the 1930s was exactly the same as Ibn Batuta described it six centuries before. Naked except for a loin-cloth and leather protection over his fingers, the diver descends rapidly to the bottom on a heavy stone which is attached by rope to the boat. His nostrils are stopped by a wooden clip on his nose; he carries a knife to cut the oysters from the bottom, working as fast as he can while his breath lasts; tied round his neck is a small basket to carry his catch. The haulers in the boat above pull up the stone after each dive, and the diver surfaces unaided when he can stay down no

29. Raising the ceremonial banner of old Kuwait, with a regiment of Model-T Fords in the background

longer. Divers used to work for six to eight hours a day, with only short periods for rest. Each batch of oysters brought from the sea bed was thrown on deck in a communal heap, where the shells remained untouched till evening or the following morning when all hands working together would open them to see what they might yield.

Like all the Arabs' seafaring activity, pearl-diving was carried on according to ancient rules and rituals. Haulers and divers remained always in their distinct groups, and neither did the work of the other, and the rule of never opening an oyster till the appointed time was always faithfully observed. Another of their customs was to have a copy of the Quran wrapped in a cloth and hung from the awning struts at the stern of the boat, from where it would be taken down in the evenings so that the captain might read it to the crew.

Although the pearl trade was in decline in the 1930s it was still an activity in which practically every family in Kuwait had a stake, and its traditional importance to the town was shown by the fact that the Shaikh himself sailed down to the pearl-banks at the end of the season and gave the signal for the fleet to return to port. Then every boat would hoist sail and make for Kuwait with all speed, while the wives whose men had been absent for four months would hurry down to the seafront in their best clothes and welcome the boats home with drums and waving banners.

Ikhwan rebellion

Between 1927 and 1930, during the reign of Ahmad al Jabir in Kuwait, Ibn Saud's kingdom was beset by a wave of anarchy and revolt which came near to overthrowing Saudi power. During these years the Ikhwan, whose movement the King had fostered for his own ends, broke free of his control, and in their final active rebellion were only subdued because Kuwait refused to allow them to take refuge in her lands, and the planes of the RAF from Iraq were used to force them to surrender.

The grievances of the Ikhwan stemmed from a fundamental difference in outlook between themselves and Ibn Saud, a difference which had remained latent while they were engaged in warfare, but which came into the open after the Hejaz campaign of 1925. In that year, Faisal al Duwish had captured Medina and forced the abdication of the Sharif Hussain. The Asir had already been subdued, and Ibn Saud now controlled the greater part of the Arabian peninsula. But now that there were no more lands to conquer, the fierce beduin warriors who had gone into battle with such zest, were not going to be content to return quietly to their tents and the pastoral life. After the fall of Jedda, the Ikhwan leaders had been bitterly disappointed that Ibn Saud had not allowed them to plunder the Hejaz, and they were further angered by his prohibition on raiding northwards across the frontier. It was obvious that he aimed to curb their taste for warfare and impose orderly government in his territory. Here lay the main point of difference between the Ikhwan and the King. They were bedu pure and simple, living in the way that had been traditional since time immemorial in Arabia; they wished to continue in this way, remaining free and independent of all but tribal authority. Ibn Saud, on the other hand, while he had affected to share their beduin ideas and fanaticism when it suited him, was a shrewd politician. Finding himself in possession of most of Arabia, he realised that to keep his kingdom he must move with the times in a changing world. To maintain his prestige outside Arabia, it was essential not only to create an effective central administration, but to show that he could impose law and order on his turbulent tribes. The first step in this direction was to ensure the safety of the desert caravan routes. To this end, he forbade all raiding and looting, and began to set up a network of radio communications to enable him to enforce his authority throughout his wide domains. He also equipped himself with motor transport which could move faster than bedu on camels to any possible trouble spot.

Raiding and plunder, as we have seen, were the breath of life to the bedu – activities in which they proved their manhood, satisfied their need for action and exitement, and increased their material wealth by stealing camels. The King's new policy ran counter to ancient desert traditions.

There had been a time when in the eyes of the Ikhwan, Ibn Saud had personified all the true beduin virtues, but now they began to feel acute disillusionment. They foresaw that the character of Arabia was changing and they resented change. In addition to his pacification policy, they hated his introduction of western inventions such as cars and wireless, which were anathema to their narrow religious views. Many bedu felt that by adopting such innovations (which by their lights were contrary to Quranic law) he was giving in to British pressure.

Faisal al Duwish, who all along had fought his campaigns in a spirit of reforming zeal, believing that he was helping to reassert the power of true religion over people who had fallen into evil ways, now saw that his friend and leader had manipulated the Ikhwan to suit his own political needs. Feeling that he and his own tribe, the Mutair, could no longer take orders from a ruler whose religious sincerity was in doubt, he conferred with two other leading tribal chiefs, Ibn Humaid of the Ataiba and Dhaidan al Hithlain of the Ajman. Together they made a pact to support each other against Ibn Saud if he should act against any one of them.

At this time, in 1927, Faisal was actively spreading propaganda against the King, and throughout the Ikhwan movement there was very evident discontent with Ibn Saud's government, and his conciliatory dealings with Britain and other countries. When Imperial Airways requested permission to use the eastern seaboard of Arabia as a landing point on their proposed route to India, Ikhwan hatred of western technological inventions was so strong that Ibn Saud felt bound to refuse.

Before long a new cause of contention arose in the north. Iraq built a police post in its southern territory at Busaiya, and although this was fifty-five miles from the Neutral Zone frontier, the Ikhwan saw it as a threat to themselves. They demanded that the King should take action against both this post and another at Salman; they were unsatisfied with efforts he had made through normal diplomatic channels. The Ikhwan believed that Iraq – always regarded by them as an infidel country – was actively plotting with Britain to oppose them, and Duwish openly declared that the existence of the Busaiya post was proof of Ibn Saud's weakness in the face of provocation, and his inability to stand up to Britain and Iraq.

In November 1927, on Duwish's orders, the Mutair attacked the Busaiya post and killed all but one of its occupants. Among the dead was a woman. At this time Ikhwan ferocity and fanaticism were at such a pitch that they even abandoned the ancient principle of women's sanctity in warfare.

The attack on Busaiya has always been regarded as the first active sign that the Ikhwan had broken away from Ibn Saud's authority, but underlying this action was a mystery that has never been clarified. Faisal had given his followers the command to attack the post, but he himself always insisted – and to this day the Mutair believe this – that the orders came first from Ibn Saud. The attack into Iraq brought the RAF from Shuaiba into the drama which was developing between the Ikhwan and the King. It is more than possible that Ibn Saud had given his approval for the attack, knowing that it would provoke British reprisals against the Ikhwan fanatics and aiming to achieve just this. The incident marked the beginning of two years of strife and turmoil in the desert.

After the Busaiya raid, Duwish's Mutair tribesmen flouted the King's authority in raiding and looting at will. They came into Kuwait territory and attacked Kuwait bedu at Umm Rimmam north of Jahra during November 1927, and in Kuwait town a state of constant alarm existed. The town wall was repaired and manned nightly until the end of that year. In addition, the Jahra garrison was strengthened to three hundred men.

In early December, Ibn Saud informed the High Commissioner in Iraq and the Shaikh of Kuwait that the Ikhwan, contrary to his orders, were marching north to attack Iraq. The following day RAF planes over the Neutral Zone located and machine-gunned Ikhwan raiding parties. This time it was clear that Faisal al Duwish was leading his tribesmen against Iraq in defiance of the King's orders.

When, in January 1928, another raid was made into Kuwait territory by Ibn Ashwan, who attacked Umm Ruwaisat thirty-eight miles north-west of Jahra, Shaikh Ahmad determined to act against the raiders. All available cars in Kuwait were commandeered – twenty-five were mustered – and went to Jahra the same evening, filled with armed men. Their instructions were to proceed to Riqai (the point at which the Iraq, Najd and Kuwait frontiers join) at first light to cut off the raiders' retreat. Travelling from Jahra over rough country with no road, the Kuwait force succeeded in getting fifteen of its cars to Riqai, where they intercepted the raiders and attacked. This was the first desert battle in which motor vehicles were used, and it was largely due to this new factor that the Kuwaitis were able to defeat the Ikhwan force and make them abandon a large quantity of booty. In this engagement, Shaikh Ali al Khalifah (a second cousin of Shaikh Ahmad), who had commanded the Kuwait force, was severely wounded. The total Kuwait fatalities numbered eleven, while the Mutair lost over thirty-five men. Shaikh Ali al Salim al Sabah, arriving late at the battlefield because of trouble with his car, insisted on pursuing the enemy on his own, and was later killed when he caught up with them in the Batin.

By 1928, there was danger of a general uprising of Najd tribes under Faisal al Duwish and Ibn Humaid, who were together demanding that Ibn Saud should declare a jihad or holy war against Iraq. The Ajman, who had never shown any enthusiasm for the Ikhwan movement when it was the tool of the King, readily joined the movement now that it appeared to be acting independently. Being beduin born and bred Ibn Saud understood perfectly the unformulated hatreds which were brewing among his tribesmen, and did his best to express these to Sir Gilbert Clayton at the Jedda Conference in May 1928 where the question of the Iraqi police posts was discussed. No agreement was reached. Ibn Saud was in a dilemma, caught between tribal demands and a genuine desire to appease Britain.

So widespread was the general dissatisfaction with the King's leadership that at this time Ibn Saud dared not take action against the three powerful insurgent tribes, fearing that he could not command enough support to defeat them. But in September, he summoned a conference of all leading Najdis and tribal shaikhs in an attempt to regain his people's confidence. Although the three leading dissident chiefs refused to attend, other Ikhwan leaders put their grievances to the King, whose promises to them went some way to allaying their suspicions. By the end of the conference, Ibn Saud believed that he had recovered enough support to be able to act effectively against the Mutair, Ataiba and Ajman.

30. Faisal al Duwish, Paramount Shaikh of the Mutair. Sketch made by the late Squadron-Leader H. Stewart, R.A.F.

At the beginning of 1929, general lawlessness existed throughout the Kuwait hinterland. The Iraqi shepherd tribes were on their annual migration to Kuwait territory, and they were the main victims of Ikhwan attack. In January, during a raid against these tribes, Dhaidan al Hithlain's Ajman tribesmen shot and killed an American missionary who was driving to Kuwait along the road from Basra. After this incident the Basra road was closed for a month owing to the general danger from marauding Ikhwan bands. The most serious of these attacks on the Muntafiq tribes in Kuwait occurred at Jalib as Shuyukh, barely seven miles from the town. Here, six hundred men descended at dawn on forty tents of Bani Malik. Thirty-seven Iraqis were killed and six thousand of their sheep and six hundred and fifty donkeys driven away. Thirty families had their tents and all their possessions stolen. The raiders were pursued by the RAF but escaped with their loot. Many destitute survivors had to be repatriated by Kuwait to Iraq.

During March the King took command of his forces and left Riyadh for Zilfi, where he called upon Duwish and Ibn Humaid to meet him and negotiate. Faisal al Duwish dined with the King in his tent, and left with the understanding that he and the King would continue their discussions on the morrow. Ibn Humaid had camped nearby, but had refused to come to Ibn Saud's tent. Contrary to all laws of beduin hospitality, Ibn Saud ordered his army into battle at Sibila at dawn the next day. Faisal al Duwish was surprised in his tent and seriously wounded in the stomach without having taken part in the fighting. The Ikhwan were defeated, and though Ibn Humaid escaped from the battle he surrendered soon afterwards and was imprisoned. To the beduin world, Ibn Saud had overcome the Ikhwan forces by an unforgivable act of treachery, breaking the unwritten law by which a man who has taken food in a tent is regarded as safe from attack from that tent for three days afterwards.

The King, believing that Faisal al Duwish was dying, sent him back to Irtawiyah, and considered that the Ikhwan had now been crushed. But Dhaidan al Hithlain was still free. Though he had stayed away from Zilfi, he fell victim to another act of treachery committed by Fahad Ibn Jiluwi, son of Ibn Saud's governor of Hasa. He had gone to visit Fahad after being promised safe conduct, but was killed while he was a guest in Fahad's camp. Although Ibn Saud was not directly responsible for Hithlain's death, this act roused further beduin sentiment against the King and his supporters throughout north-east Arabia.

While everyone believed that Faisal al Duwish was dying, he was in fact making a surprise recovery from his wound as he lay low at Irtawiyah. While he was out of the fighting, the Ajman, whose full fury had been roused by the treacherous murder of their chief, now openly declared their intention of overthrowing the King. They were supported by other Ikhwan chiefs, and soon were in control of the whole of northern Najd between Jabal Shammar and the Gulf.

Britain, who had supported and subsidised Ibn Saud in his rise to power, had a vested interest in maintaining the Saudi regime which seemed to offer the only hope of stability in Arabia. Through the Political Agent, Kuwait was informed that since the Ajman were in rebellion they must not be allowed to take refuge in Kuwait, nor buy supplies in the town. Shaikh Ahmad, valuing his friendship with Britain, pledged his word that he would give no asylum or assistance to the rebels.

In May 1929, Faisal al Duwish, restored to health, re-emerged as leader of his tribe and joined the Ajman in active revolt. Having gathered a force of five thousand men and a hundred thousand camels from his own tribe and the Ajman and Ataiba, he moved to Jariya Ilya, just south of Kuwait. Formidable though his huge force was, it was going to create supply problems which in the end would prove insoluble. Ibn Saud probably foresaw this, and refrained from taking immediate action.

Meanwhile, from Jariya, Faisal invited Shaikh Ahmad of Kuwait to join the rebels and recover some of Kuwait's former lands. He requested permission to enter Kuwait and camp at Subaihiyah wells. But the British government had threatened action by the RAF from Shuaiba against rebels who might cross into Kuwait territory. During the following weeks Duwish wrote more than once to the Shaikh, calling on him to act as his great ancestor Mubarak would have done. In former times, the Ajman, Awazim and Mutair had all paid allegiance to Kuwait, and here was a chance, said Faisal, for Ahmad to recover not only his former territory but also his overlordship of these tribes. It must have been a tempting proposition to the Kuwait ruler, but he stood firm on his promises to the British government. If Kuwait had joined the rebels, and British forces had remained neutral, it is probable that Ibn Saud would have been overthrown, since the Najd fighting forces were in sympathy with the rebels.

By August, the insurrection was gaining in strength, after several successful rebel attacks against the King's forces. But the need to find fresh grazing for their vast numbers of camels, as well as food and ammunition for themselves and their families, led the Ikhwan once again to enter Kuwait. At the end of August, Faisal and his supporters arrived at Subaihiyah. Dickson, then Political Agent in Kuwait, met Duwish personally to deliver an ultimatum that they would be bombed unless they left within forty-eight hours. Duwish argued that he had no quarrel with the British, and that as his tribe were old subjects of the Shaikh of Kuwait they wished to return to their former allegiance. His army was short of rations and he urgently needed to buy food in Kuwait. However, on Dickson's insistence, the rebels withdrew.

When the worst of summer was over, it was apparent that the King was making preparations to go into action. In October, while Faisal engaged and defeated the loyal Awazim in Hasa, the King's forces in western Najd won a decisive victory against the Ataiba. This was the severest blow the rebels had yet suffered. On 30 October, Duwish arrived near Jahra, asking to see Shaikh Ahmad and Dickson. He was told to withdraw immediately, but in a letter to Dickson, he asked whether, if he left the women and children and camels of his tribe in Kuwait, they would be allowed to remain under the Shaikh's protection. Dickson was instructed to give him no encouragement. This meant that he could not leave to attack the King in Najd, since his women and camels would be at the mercy of the Iraq tribes while he was absent. The Mutair women were starving and reaching the end of their endurance, for in beduin fashion they had moved with the Ikhwan forces from the beginning and had been unable to buy food since Kuwait had forbidden them facilities.

From this point, Faisal al Duwish appeared to lose heart for the fight. He had suffered a bitter blow the previous August, when his favourite son Azaiyiz had died of thirst in the desert after being cut off from water while making a raid north of Hail. Now it seemed the weight of his problems was more than he could bear. His companions noted a change of character in their chief, who began to advise them that any who wished might leave and make their peace separately with the King.

By December, when his forces were much reduced in numbers, Faisal al Duwish, now in the Batin, decided to ask for surrender terms from Ibn Saud, who was encamped with his army at As Safa. RAF ground forces in armoured cars were concentrated at Riqai to prevent the rebels from crossing into Iraq or Kuwait. It appeared that Duwish and his men were cornered unless he could find asylum in Kuwait. At this point Dickson, genuinely concerned at the plight of the rebels, disguised himself as a beduin and made his way out to the camp of Naif al Hithlain the Ajman leader, to advise him to surrender, since there was little hope for the Ikhwan against the Saudi forces, and the unfortunate women and children of the rebels would face yet more privation and suffering unless hostilities were quickly ended. Hithlain refused to consider surrender.

The appearance of Ibn Saud's army at Riqai in early January 1930 forced the rebels to move back to Jahra. On this march, Dickson, with some of Shaikh Ahmad's guards, tagged along among the rebel bedu, marching through the night with them and later sending word to the RAF giving their new location. To those who do not know the desert it may seem strange that an army can disappear in a featureless landscape, but it is a fact that once men or animals have passed beyond the horizon they can rapidly go off in an unexpected direction, and without air observation, it may take days to find them.

RAF planes and armoured cars overtook the rebels ten miles west of Jahra, in a state of total disorganisation. By 7 January the main body arrived at Jahra, where they were pinned down by intermittent bombing all round the village. Ibn Saud's army, concentrated on the southern frontier, was waiting to annihilate the rebels if – as he expected – they were forced to leave Kuwait. Dickson, returning to Kuwait, went back to Jahra by car, running the gauntlet of machine-gun fire to enter the village. He pleaded unsuccessfully with Duwish to surrender rather than make a bid to escape and meet certain disaster against the Saudi army. In the confusion which prevailed in Jahra during the bombing, Faisal al Shiblan, head of the Jiblan clan of Mutair, and Duwish's second-in-command, slipped out of the oasis. Among the bedu it was thought he might claim the right of *dakhala* or sanctuary with Shaikh Ahmad of Kuwait, a situation which would cause tricky diplomatic problems, since by sacred Arab custom the Shaikh would then be bound to protect him.

Through Dickson's persuasion, Naif al Hithlain, the Ajman leader, finally agreed to surrender to the RAF on 9 January. Duwish, who until then had hoped to fight his way out of Kuwait and back to Najd, now saw that capitulation to the British was the only course. He and Sahud Ibn Lami, chief of the Jiblan clan, handed over their swords to the Chief of the RAF Air Staff, Iraq (the late Air Vice Marshal Sir Charles Stuart Burnett). To the Mutair, this was almost unbearable humiliation, and to the whole Ikhwan fraternity a moment of ignominious defeat.

But by surrendering to the British, the leaders escaped the sentence of death which almost certainly would have been their lot with Ibn Saud. The three chieftains, proud and stoical in the face of disaster, were flown to Basra by the RAF and taken aboard HMS *Lupin* lying in the Shatt al Arab. The remnants of the Mutair and Ajman, with their camels, were kept under guard by the RAF until they could be handed over to the Saudi King.

Before Faisal al Duwish left for Basra, he had entrusted his Dushan womenfolk to Dickson's keeping, and the Englishman, understanding the bedu and their ethical code, deemed it a privilege to have these women in his care. Thirty-seven of them were accommodated in the Red Fort and looked after by Dickson and his wife until lodging could be found in Kuwait. The Dicksons' kindly help earned them the profound gratitude of the whole tribe, and later on the personal thanks of Ibn Saud himself.

A day or two after the surrender, Dickson was in his office in Kuwait, when a servant announced the arrival of a strange beduin shaikh. A few minutes later a tall impressive figure, wearing the white turban of the Ikhwan, was ushered into the room. The Political Agent, who knew most of the chieftains of this part of Arabia, did not recognise the man. However, he uttered the customary greetings and called for coffee to be served in the usual way.

After drinking coffee, the visitor informed Dickson that he was Faisal ibn Shiblan, Duwish's second-in-command who had escaped from Jahra, and had come to consult the British Political Agent on his next move. He wanted to know if British pressure would force Shaikh Ahmad to deliver him to the RAF if he appealed to the Kuwait ruler for sanctuary. If Dickson advised him that Shaikh Ahmad could not protect him he was ready to try to escape to Syria.

Dickson, filled with admiration for the man's courage and frankness, advised him not to make *dakhala* with the Kuwaiti ruler, knowing this would create an extremely difficult situation for both the Shaikh and himself. Instead, backing his intuitive judgment of the beduin character, and his personal knowledge of King Ibn Saud,

Dickson advised Faisal to go straight to the Saudi King and make his *dakhala* there. He believed the King would respond by granting sanctuary and pardon.

While Ibn Shiblan was pondering Dickson's advice, the office boy entered with a telegram. It was from the AOC at Jahra. Burnett had been reliably informed that Ibn Shiblan was in Kuwait, and he now asked Dickson to request Shaikh Ahmad's help in finding and arresting the rebel and sending him in irons to the RAF at Jahra.

Dickson proceeded to translate the contents of the telegram verbatim to Ibn Shiblan, who listened with admirable self-control.

"What do you think I should do?" asked Dickson.

"You know best," Ibn Shiblan replied simply. Dickson was won over by his composure and lack of fear and told the wanted rebel chief that he knew and honoured the Arab code, and since Faisal had drunk his coffee he was safe. Dickson directed him to the home of a Mutairi merchant where he might lie up for a few hours. He then arranged that the merchant should have a fast riding camel waiting outside the town wall at midnight, so that Faisal, helped over the wall by a rope, could make his escape on the camel. Faisal for his part said that he would take Dickson's advice and throw himself on the mercy of the King. Leaving nothing to chance Dickson told him to reach the Saudi camp at nightfall and enter the King's tent after dark. His chances of survival would be less if he encountered anyone but the King himself.

Faisal made his escape according to plan, and in the end Dickson's assessment of the King's reaction was abundantly justified; Ibn Shiblan was forgiven and taken back into the King's service, escaping the imprisonment which the other leaders ultimately suffered. As for the AOC – by Dickson's account he never mentioned the matter again.

Meanwhile, the business of handing over the remaining rebels, with all the camel-wealth of their tribe, had still to be discussed with Ibn Saud. To this end, the Political Resident in the Gulf, Col. Hugh Biscoe, and Air Vice Marshal Burnett arrived in Kuwait, and a rendezvous with the Saudi King was arranged at Khabari Wadha, ninety miles south-west of Kuwait. Biscoe, Burnett and Dickson flew there on 20 January 1930.

After some hard bargaining terms were agreed with the King in return for the handing over to him of Faisal al Duwish, Naif al Hithlain and Ibn Lami.

The rebels' lives would be spared. Any punishment would be tempered with mercy though he reserved the right to recover loot taken by the rebels from others. He would prevent further raids on Iraq and Kuwait. He promised to settle part claims for restoration of loot from Iraq and Kuwait, provided all rebels and their camels now in Kuwait territory under guard by the RAF were returned to him. He agreed to pay £10,000 in compensation to Iraqi and Kuwaiti tribes, in anticipation of a final settlement of accounts.

The King was prepared to name representatives to a tribunal, to be set up according to the 1925 Treaty of Bahra between himself and Iraq, to discuss claims against each other by Iraqi and Najdi tribes.

The British mission returned on 27 January with the agreed terms signed. The next day, the three Ikhwan chiefs were landed from *Lupin* and taken by air to Khabari Wadha, where there was a touching scene of reunion with the King. Such are the paradoxes of beduin character that men who had fought each other fiercely could still feel the bond of friendship and shared experience of former times despite the bitter enmity that had set them apart. These men were aristocrats of the tribal hierarchy, and were honoured as such.

When Dickson took the prisoners into the King's marquee he was delighted to see Faisal ibn Shiblan sitting among Ibn Saud's chiefs and advisers, and the King, noticing his glance, said: "Yes, Dickson, we have forgiven him. He came and made *dakhala* to us and told us all."

The three rebels were imprisoned by the King for life. In addition, all Duwish's personal horses and camels were forfeited. The Dushan tribe was divested of all loot acquired in raids and the Mutair as a whole lost their sacred herd of black camels, the Shurruf, as well as all riding camels, horses and mares. The Ajman received similar punishment.

The remaining body of rebels who had been kept in Kuwait territory were, despite the presence of the RAF, attacked almost nightly by the Iraqi tribes who tried to get their own back for past sufferings. On 29 January, they began the return to Najd and were handed over to the King's guards on 4 February.

For remaining neutral through the period of the rebellion and keeping his word to Britain Shaikh Ahmad, in 1930, received the KCIE. Though the British government also promised to use its best efforts to bring about the end of Ibn Saud's blockade of Kuwait, this was not achieved for another seven years. It is interesting to recall that after the Uqair settlement, Shaikh Ahmad asked Sir Percy, "If I become strong like my grandfather will the British government object if I denounce the unjust frontier line and recover my lost territories?" Sir Percy had laughingly given him his blessing. At the time of the Ikhwan rebellion he might have had a chance to do just that. Perhaps it was his Arab sense of honour that kept him loyal to his pledge to Britain.

Faisal al Duwish died in prison on 3 October 1931.

A meeting of opposites

There was a quality of inevitability about the part played by Dickson in the bitter struggles and the tragi-comic fall of the Ikhwan. The Political Agent who, with the possible exception of Shakespear, served Kuwait with greater zeal and dedication than any of the advisers sent by Britain to supervise her qualified presence, was a romantic at heart. His favourite author was Rider Haggard. Dickson would not have dreamed of ordering a suit except from Savile Row, or of wearing shoes that were not bespoke. He usually wore a flower in his buttonhole. Yet dress and taste were often at odds with habit. Much of his time in Arabia was spent in the desert. And if the visitor to the Agency was lucky enough to find him at home there was a good chance that he would be working on his car, which he maintained himself with some mechanical ingenuity. His taste for adventure was boyish, his manner – especially among the Arabs and with women – was easy and courteous. Analogy with the literary heroes of the secret service in the heyday of the Empire springs readily to mind. The comparison would do less than justice to the man, however. During his seven years as Political Agent he demonstrated a great deal more by way of skill and aptitude than an ability to cut a figure in the distant outposts of the India Office administration. His attachment to Arabia and its people was deeply genuine and it won him the confidence of rulers and ruled alike. His knowledge of the desert community, its ways and attitudes of mind, was vast and immensely valuable to Britain. His integrity was sometimes aggressively apparent and it left no doubt on either side of a dispute or a bargain that he could be trusted implicitly. On the other hand, he was not given to compromise or even to subtlety of negotiation. He distrusted businessmen almost as much as socialists and other transgressors of his rigidly Victorian belief in the inevitability of things as they stood. He admired the Arab way of life, and code of conduct. But for his own countrymen a different code of behaviour and morality was ordained, and he had little time for the man who confused one with the other.

He worked closely with Sir Percy Cox in the crucial period which saw the emergence of modern Arabia from out of the expedients of wartime bargaining, and thus he grasped threads of political ambition and intrigue that were not always apparent to some other servants of the Foreign and India Offices. He was familiar – but not always sympathetic – with other great Arabists and Arabophiles of his period, people like Philby, whose passions he understood but whose precepts were beyond him, and Gertrude Bell whom he greatly admired. Equally, he was on terms of close personal friendship with Ibn Saud and with the ruling family of Kuwait. His background was as conventional as were his ideas. St Edward's School and Wadham, the Mesopotamian campaign, the army and political service. A man of action rather than a scholar, honourable, opinionated, a keen sportsman, a good water colourist, Lt Col Harold Richard Patrick Dickson was a gentleman of the old school, with all the qualities of uprightness and prejudice that the description evokes. He was admirably suited to the task that confronted him at the time he became Political Agent in 1929, when the Ikhwan rebellion had reached its climax and Shaikh Ahmad's Kuwait came face to face with the antagonism of its neighbours and the realities of commercial negotiation.

31. Lt Col H. R. P. Dickson

His wife Violet, with whom he shared every aspect of his political, military and literary activities, contributed much to a career that, if it brought no special material rewards, gave immense satisfaction to both. They found the same pleasure in the desert; in its people and wild life, and its remoteness. They mixed among the bedu with complete ease and understanding. And while he recorded the political complexities of desert rivalry and made detailed observations of customs and habits which were to find their way into his two major books, *The Arab of the Desert* and *Kuwait and her Neighbours*, she documented the plant life of the region and showed herself to be a naturalist of no mean ability.

Dickson spoke the Arabic of the desert, though he never properly learned to read or write it. He was an inveterate story teller and his greatest delight was to talk endlessly with the bedu. The same talent was not always appreciated by his fellow countrymen. It was said of him and his wife that if they were going to see friends, it meant they were off to the desert. If they were meeting Europeans they would simply say they were seeing the Jones or the Smiths. Certainly in his later years, when he became Chief Local Representative of the

Kuwait Oil Company, he was ill at ease with Europeans who did not share his close interest in Arabia. When the Dicksons were not at camp with their *friends*, they often entertained them in the old mud-brick Political Agency to which they moved in May 1929, sharing with their guests delicacies of the desert such as dried locust.

The old agency building, one of the few traditional structures left behind by the bulldozers of the later oil era, is described by Violet Dickson in her book *Forty Years in Kuwait*:

"It was a two-storey structure, shaded along the upper floor by wide verandahs overlooking the harbour. It had been adapted from a traditional Arab-style house, built – like most others in Kuwait – with walls of sun-dried mud bricks and coated with a plaster of white gypsum. Heavy rain would soak through the plaster and soften the mud of such walls, and these houses needed constant repair after every wet season. The ground floor consisted of agency offices, post office, and kitchen, and the living quarters were upstairs. A tall arched passageway made a cool dark corridor straight through the lower storey from the main gate on the road to the inner courtyard, and the walls here were permeated with damp. The trouble was that the house stood at the bottom of a slope, and the rain-water which came from three houses on the hill behind flowed across our yard and out through the arched corridor down to the sea. The guest-room of the agency was on the upper floor of a separate block which stood a few yards from the main house. A little wooden bridge ran across from the house to the verandah of the spare room, but when we arrived this verandah was lying in a heap on the ground, and access was impossible. We found a clever Persian builder, Usta Ahmad, who came to our rescue. Using coral rocks from the sea (the only stone available in Kuwait) he strengthened the foundations and built four buttresses in the front of the house. Although it needed further attention almost every year, we made the house comfortable, and it has now been my home for forty years (apart from Harold's last year as Political Agent, when we moved into a new Agency). In the old days it had two curving stairways leading down to the road from the entrance to the upstairs living quarters, and these, with its green-painted woodwork, gave it a distinctive look, but in general style it blended in with the houses flanking it on either side along the sea-front. Today, this same house is one of the very few of its period which remain in oil-rich Kuwait. All the pleasant old mud-brick houses which formerly surrounded us on three sides have been bull-dozed away, and replaced by characterless structures in reinforced concrete. My house is almost a museum piece, only preserved because I have been allowed to stay there. For the continued use of this 'Grace and Favour' residence I have to thank the Al Sabah family, and the unfailing kindness of the ruler of Kuwait, whose property it is."

32. The old Political Agency as it appeared in 1970

Mrs Dickson also describes the contrasting scenes of the town as she knew it then, and in later years:

"Often in the hottest weather, if there was hard work to be done on the boats along the seafront, the sailors would try to do it in the cooler night hours, and then we would hear the rhythmical songs which helped them to heave together to raise a heavy mast, or turn a capstan to drag a boat ashore. Then at first light, if we were not too sound asleep, we could hear the call of the muezzin from the small minaret of a neighbouring mosque – the authentic voice of Arabia, a resonant but simple call made by an old blind man. But all these things have gone. Today, elaborate new minarets have replaced the unpretentious ones of former times, and the call to prayer, alas, booms from a loud-speaker. . . . Once, at about the time when we were settling down for the night on our roof, there was an eclipse of the moon. In those early days the Arabs were much alarmed by such unusual happenings, and on that occasion I remember bands of drummers and singing men roamed the streets, praying and calling for the restoration of the moon's light."

When Col Dickson took himself off into the desert, sometimes alone, sometimes with his family, conversation was of a politeness both impeccable and touchingly simple.

The beduin host would usually ask after the weather in England, though Dickson's home visits were few and far between.

"Rain has fallen daily through the past two months," he would say, "and the grass stands waist high in the fields."

"Truly God is great," the host would reply, "and England fortunate; the flocks and herds of England must be fat and her people prosperous."

The bedu would ruminate on the joy of relentless rain. Then the company would settle to a meal and idle chatter. It was important to know the state of prosperity of one's host at such times. Such is the Arab's sense of hospitality, he would be liable to serve up his one and only sheep if a visitor had to be entertained and there was no other food. Sometimes Dickson would regale his friends with stories of mechanical contrivances in urban civilisation. The London underground, for example.

"How is it possible for these things full of people to go underground?" the bedu would ask in wide-eyed wonder.

"Well, these things are like the jerboa of the desert," replied the resourceful Colonel. "As the jerboa digs its burrow in the ground it runs back and forth through the tunnel it makes. So do these trains."

Many years later, when the second world war was over, he visited Salim, an old beduin acquaintance. Salim had heard about the war and about someone called Hitrel, a German. He was assured that the war was over and that Mr "Hitrel" had taken his own life at the moment of defeat.

"God is just. And all-knowing", said Salim.

To a man like Dickson who loved the bedu and spoke their language, the most rewarding part of his work as Political Agent was talking to the tribesmen and gleaning from them intelligence about affairs and opinions in Arabia as a whole. Movements of tribes, local enmities or alliances, beduin criticisms of the central Saudi or Kuwait authorities, all made up valuable information to be passed on to the Political Resident in the Gulf and duly weighed in formulating British policy in Arabia and the Gulf states. He was at home in the desert, a devoted friend of the Arab and an untiring servant of Kuwait.

Another European personality who played an important part in the history and transformation of Kuwait was Major Frank Holmes.

A New Zealander by birth, a mining engineer by training, he had already covered most of the old world and much of the new in search of mineral wealth and personal prosperity when the 1914–18 war broke out.

At the start of that war he joined the Royal Marines. Later he became Supplies Officer to the Royal Naval Division, an organisation formed at the instigation of Winston Churchill.

Competition for places in this force was fierce. Frank Holmes, never one to waste time with lesser fry, contrived an interview with Churchill himself, then First Lord of the Admiralty. The meeting was characteristic of both men. It is related in Wayne Mineau's well-informed if strangely titled book, *The Go Devils*.

"How old are you?" asked Churchill.

"Just forty sir," said Holmes.

"That's a bit old, isn't it?" said the First Lord.

"It's exactly your age, sir," said Holmes.

"In that case you're in the prime of life. We'll see what can be done," came the reply.

It was as supplies officer to the Naval Division that Holmes gained his first view of the Arabian desert. While others engaged in political and military warfare, pursuing the leaders of Arab countries in the Allied cause, Holmes wandered nonchalantly from one country to another looking for meat for the troops. He also engaged in active combat.

The war over, he joined with a group of engineers in 1920 to form Eastern and General Syndicate, with the object of acquiring oil concessions in the Middle East and selling them to countries able to undertake the work of exploration and operation. He left England for Arabia in 1921. As we have seen he turned up a year later at the Conference of Uqair.

A month after the Dicksons arrived in Bahrain, where they awaited Sir Percy Cox, they were told that a certain Major Frank Holmes and a colleague, Dr Mann, were due to arrive there. They were asked to provide board and lodging.

Dickson's suspicions were aroused. He had only recently heard of oil seepages near the Qatif oasis on the mainland, and had been as far afield as Dhahran by donkey and camel in search of them, but still he could find no evidence to back the assertions of bedu tribesmen that they had seen oil in the desert. On Tarut island, he was shown a copy of a Turkish report which stated that an oil seepage existed behind Qatif. He told Sir Percy of his suspicions regarding the visit of Holmes.

The Dicksons awaited the arrival of their guests with some interest. They were not prepared for the actuality, however.

"Two things greatly amused my wife and myself. The first was the appearance of Major Holmes. He carried a large white umbrella lined green, wore a white helmet as issued to French troops in Africa, and over his face and helmet a green gauze veil – quite like pictures one has seen of the tourist about to visit the pyramids. The second was the amazing number of presents Holmes had brought for Ibn

Saud. There must have been over fifty cases, leather bags, boxes and guns. The two new arrivals very soon made themselves at home. Major Holmes was an amazingly amusing companion. His anecdotes were legion. He exuded charm, but I soon discovered he was a sick man, having bladder trouble. He wondered how he was going to do the journey by camel to Hufuf and possibly beyond. I was able to give him useful tips, for I had done the journey and many more while Political Agent in Bahrain."

When the time came for the visitors to set out, Holmes announced that he would be taking a camel to Qatif while Dr Mann would travel via Uqair.

"But you are a sick man. Why ride a hundred miles by camel when, by the Uqair route, you will only have to cover a bare fifty?" Dickson expostulated.

The rest of the conversation must be recounted in Dickson's own words:

"I am a butterfly collector," said Holmes, "and I have been told that a wonderful black variety, known nowhere else in the world, is to be found in the Qatif oasis. I have already called it the Black Admiral of Qatif and am out to get a specimen. Then my name will be famous."

He was going on with his bacon and eggs, when my wife dropped her bombshell.

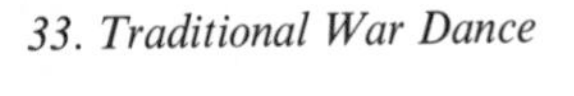

33. Traditional War Dance

"Major Holmes", she remarked in a quiet voice, "this is the first time I've heard of oil seepage being called by the name of a butterfly."

The effect was immediate. Holmes jumped up from his chair saying: "What on earth do you mean, Mrs Dickson?"

"Exactly what I have said."

He was so taken aback that he rushed round the table and clasped my wife's hand.

"My God, you are a wonderful woman!" he exclaimed. "I shall telegraph today to the curator of the Zoological Gardens in London and ask that you be made a Fellow of the Zoo."

He was as good as his word. Shortly afterwards, my wife heard that she had been duly elected FZS.

As it turned out, it was Holmes who was making for Uqair, complete with gifts for Ibn Saud.

The Dicksons were amused by their first encounter with the ebullient prospector. Both men had immense charm in their very different ways and both could tell a good story. The amusement was soon tempered by wariness, however. Dickson, with his keen sense of propriety, was not greatly pleased to see the uninvited Holmes camped between the two factions at Uqair, puffing at an endless chain of cigarettes and throwing an oily spanner into their political deliberations. When they met subsequently in Kuwait, the Political Agent was on his guard.

Date gardens in dispute

Among the countless overlapping disputes and problems of the twenties and thirties, none was more tortuous or intractable than the long legal quibble over the Sabah date gardens in Iraq. The issue may, at a distance, seem less than vital to the interests of Kuwait and the political stability of the region. In fact, it was closely bound up with the fortunes of many thousands of people in Kuwait, who often relied on these plantations for sustenance as well as luxury, and with large issues of politics. In 1930 the acting High Commissioner in Baghdad had suggested to the government of the day that Britain should encourage the gradual absorption of Kuwait into Iraq. The Foreign Office, however, restated the intention of maintaining Kuwait's independence and of honouring its treaty obligations. In a sense, the date gardens represented the final aggravation in a situation which was kept on the boil by the recently-ended Saudi blockade, the customs problem with Iraq, the incursion of Iraqis bent on trouble making and the multiple loyalties and obligations of Britain with regard to all three disputants. That situation was not helped by the contention of one or two representatives of HM Government in the area that Kuwait was a small and expendable state which could be sacrificed without too much concern if the power struggles of the period demanded it.

When the subject of these gardens was first aired in the Iraq courts, the lawyers on both sides had only the haziest idea of the real facts of ownership or of the disposition of the lands in question. It could hardly have been otherwise since most of the gardens were purchased and resold under conditions of land registration which had more to do with Ottoman expediency than legal definition. Some, indeed, had not been registered at all because of Turkish nationality demands. His Majesty's nonplussed law officers, helped by the Political Resident and Political Agent of the day, endeavoured to establish the facts some three years after litigation had begun its passage through the Iraqi courts.

In giving effect to its promise of 1914, Britain had made separate assertions to Shaikh Mubarak and to the Shaikh of Muhammerah. To the former, in a letter dated 3 November, the English version of the Political Agent's arabic read in its essential part: ". . . your gardens between Fao and Qurnah shall remain in your possession and in the possession of your descendants, without being subject to payment of taxes or imposts." A similar letter to Shaikh Khazal, whom Britain also wished to reward for wartime assistance against the Turks, was couched by Col Knox in slightly different terms. It read: ". . . the date gardens you possess on the Turkish side of the Shatt al Arab shall remain in full possession of you and your heirs and immune from taxation."

In March 1933, the law officers decided that the wording of these promises was deficient in some respects. Did "immunity", or "without being subject to payment of taxes or imposts", relate to the gardens themselves or to their produce? Did the promises rely on the remoteness of the owners? Supposing either shaikh decided to reside in one of his gardens, would he then be liable to taxation? Such were the questions asked by the law officers of Whitehall in 1933. Their report, dated 29 March, concluded that an obligation rested on HM Government to compensate the shaikhs for any taxes imposed on them. They did not hold out much hope of achieving a result favourable to Britain or the shaikhs through court proceedings. Unfortunately, matters had gone too far to prevent such proceedings. Two cases had gone before the Court of First Instance at Basra, and both were now on a merry-go-round of appeal.

A description of the properties was put to paper by Col Dickson in September 1930, when litigation began. The main estates were, he said, those of Faddaghiyah and Fao, stretching northward from the former district. At that time these were worth £80,000 and £300,000 respectively. The other estates he named were Farhawiyah (valued at £1,000), Mutaawah (£22,000) and Ajairawiyah (£5,000). Within these areas, usually referred to as the district of Fao, five gardens were said to belong jointly to Shaikh Ahmad and three members of his family, while the remaining two gardens were owned by some 100 to 120 members of the Al Sabah. The ruler told Dickson that the estates were the lifeblood of his family and two-thirds of the population, supplying them with dates, citrus fruits, wheat, barley, lucerne, firewood, mats, and other articles.

That description was later supplemented by an India Office report, based on information supplied by Sir Henry Dobbs, resident land commissioner at Basra in 1915, which listed the Al Sabah properties as:

1. Al Farhawiyah, purchased by Shaikh Mubarak in about 1870 before "tapu" or land registration was introduced.
2. Al Mutaawah, purchased by Shaikh Mubarak at an unspecified date, covering about 1,200 acres, valued at £87,000.
3. Al Faddaghiyah, purchased by Shaikh Mubarak in 1908 from the wife of Ahmad Pasha. The Turkish authorites refused to register

this possession unless the Shaikh agreed to become a Turkish subject. This he refused to do and the property was not registered until 1915 when the British revenue commission at Basra entered it in the name of the ruler. The total area of this estate was said to be about 2,200 acres and its value was estimated at £150,000.

4. Al Bashiyah (not mentioned in the Dickson report), purchased in 1908 from Shafiqah, the widow of Ali Pasha, who sold it on behalf of herself and her children. In 1915 the British revenue commission at Basra asked the courts to examine the status of these gardens and to regularise the matter of ownership. The widow died at about that time, leaving five children. This land, which was at the centre of the dispute, was never registered, according to the India Office. It was sometimes included as part of the Faddaghiyah estate.
5. Al Fao, acquired by deed of gift from Rashid al Saadun. Covering approximately 16,800 acres, these gardens were said to be worth over £1 million.
6. Al Ajairawiyah, a small estate of 120 acres purchased by Shaikh Mubarak in 1912 and registered in the name of his daughter Sharifah. She married his nephew, Saud, who was resident in Iraq and the Shaikh gave her the garden as a wedding present. On her death, without issue in 1920, half the property reverted to the descendants of Mubarak. Saud then married Mariam, another daughter of the ruler, who inherited the other half. The title deeds to this property were lost at one stage and re-registered in the name of Sharifah.

For several years Iraq had tried to levy taxes on these gardens despite Britain's 1914 pledge, repeated in subsequent treaties with Iraq, that no tax would ever be exacted.

In August 1930, Rafiyah Khanum, daughter of Ali Pasha al Zuhair, filed a suit against Shaikh Ahmad al Jabir claiming that she was the rightful owner of one-tenth of the garden of Bashiyah. She sought, in effect, an injunction to restrain the Shaikh from interfering with her possession.

Papers were served on the Shaikh's agent in Basra, Abdul Aziz al Salim al Badr, who was said to have witnessed the purchase of the lands by Shaikh Mubarak from the widow Shafiqah and the minor children of Ali Pasha (of whom the plaintiff was one) in 1908. That, in fact, was the gravamen of the defence. The plaintiff pleaded in reply that the land in question was miri, or crown, property and that it could not be sold, and that, alternatively, under Turkish law property could not be sold by or on behalf of minor children without the contract being witnessed by a Qadhi, a religious legal functionary. Surprisingly, after the case had dragged on for several months, the prosecution withdrew and judgment was given in default for Shaikh Ahmad. A British judge, Mr Hooper, presided. Then in May 1931, the plaintiff successfully applied for permission to proceed with the case. Again, in July, judgment was in favour of the defendant, Shaikh Ahmad, represented as before by the lawyer Muhammad Effendi Wasfi. Then began a series of appeals and counter-appeals which revealed some of the subtleties of legal bargaining in the Iraq of this period. The prosecuting lawyers, it appeared, were engaged on a percentage-of-profits basis. If they lost the case, they received no fee. If they won, they had a share of the money gained by the action. And they had an ingenious way of ensuring ultimate success. When a case reached the appeal stage a list was always circulated to the judges of the court. If a foreigner or a dignitary was involved, or if the case was one of unusual complexity, a British judge would invariably choose to sit. But the lawyers, who were often in league with their countrymen on the bench, had a way of overcoming this inconvenience. They varied the names on the court list just sufficiently to conceal the identity of a litigant. Thus, when the plaintiff appealed in November 1931 and the case was referred back to the Basra court, the president, Judge Lloyd, was shown a list on which appeared a jumble of names, none of which seemed to have any special significance. He did not bother to sit. On 30 December 1931 the case was decided in favour of Rafiyah. Then Shaikh Ahmad appealed and this time Mr G. Alexander, who as legal adviser to the High Commissioner sat as Senior Judge of Appeal, claimed that he was kept in ignorance of the proceedings by the appellant's counsel. He was, apparently, shown a list bearing the name Ahmad Shaikh Jabir and failed to recognise it. It was said that an attempt to register these gardens before the British Land Commissioner, Sir Henry Dobbs, in 1915 failed because the vendor Shafiqah fell ill at the crucial moment. The court decided for Rafiyah and eventually the Shaikh was compelled to repurchase his family's shares in this property.

No sooner had that case reached a conclusion than a new one came before the court. This time, Abdullah al Zuhair, the son of Isa Pasha al Zuhair, one of the signatories of the original sale deed, claimed ownership of one-third of the gardens of Faddaghiyah through his aunt Ayesha. The respondents named were seventeen members, male and female, of the al Sabah, descendant from the late Shaikh Mubarak through Shaikhs Jabir, Salim and Nasir bin Mubarak; as well as the descendants of the late Shaikh Khazal of Muhammerah. Again the case turned on such issues as the validity of Britain's promises and the intentions which underlay them, the exact details of registration, and other less judicial considerations. The first deposition was made in November 1932. In the previous June, the ruler had written to the Political Agent:

"Our agent Abdul Aziz bin Salim al Badr has no doubt made you aware of the suit of Rafiyah bint Ali Pasha al Zuhair. I am displeased with the treatment I have received at the hands of the Iraq government in this connection as the palm-tree land known as Bashiyah, against which her claim is made, already belonged to the heirs of Ali Pasha and its sixth share fell to Ahmad Pasha by inheritance from his mother – Fatimah bint Jasim al Mishari – and was sold to my late grandfather by Shafiqah bint Abdul Hafidh, wife of Ali Pasha. She sold this land by a deed issued by the 'Shar' court at Basra. When the fees were being made over, the Office of Imperial Registers in Constantinople insisted that it be mentioned in the tapu documents that Shaikh Mubarak and his offspring were Turkish subjects. This he refused to permit and he was supported by Sir Percy Cox in his action.

"Faddaghiyah was bought by my late grandfather from Ahmad Pasha by power of attorney received from his wife Ayesha. This part was transferred through the tapu department after the British occupa-

To The President of the Court
of First Instance, BASRA.

Case N. 373/232

Plaintiff : ABDALLA EISA AL ZOHEIR, in his quality of heir to the estate of AYASHA BINT ABDALLA EL ZOHEIR.

Represented by : Maitre SOLEIMAN AL SHAWAF

Defendants :

HAMAD AND ABDALLA SONS OF MUBARAK AL SABBAH, SHEIKH AHMED, MAHMUD AND MARIAM SONS AND DAUGHTER OF SHEIKH GABER EL SABBAH, SHEIKHA DAUGHTER OF ABDALLA EL SABBAH, ABDALLA, FAHD, SABBAH, [illegible]ILY AND LATLAWA CHILDREN OF SHEIKH SALIM AL SABBAH, AND SABAH, BEZZAH, NOURAH AND MOUNIRAH CHILDREN OF SHEIKH NASIR AL SABBAH.

Represented by : Maitre AHMED ROUCHDY AND Maitre ABDEL GALIL BERTOC.

Note in reply to Writ dated
20th, October 1932.

May it please the Court.

The defendants have the honour to submit to the Court an objection as to its jurisdiction and thus respectfully ask it to declare its incompetence, but before going in detail in this point, we feel it our duty to make indentify the parties, and particularly the defendants.

All the defendants are the children and grand-children of Sheikh Mobarak Pasha Al Sabbah ruler of Al Kweit. One of them is H.H. Sheikh Ahmad Al Gaber Al Sabbah the present ruler and Emir of Kweit. He is the head of an independent principality under the protection of Great Britain. The regime of this protectorate is a special agreement concluded between Great Britain and the principality of Al Kweit on 3rd.November 1914. This agreement embodies Great Britain's guarantee to protect El Kweit as an independent principality. It was ruled by Sheikh Mubarak Al Sabbah who was succeeded by his grandson, H.H.Sheikh Ahmad El Gaber El Sabbah one of the defendants. Furthermore this agreement states the Great Britain shall preserve for the ruler of Al Kweit Sheikh Mubarak Al Sabbah, and for his children and grand children after him, their estates in Basra, enumerated one by one and amongst which appears the Faddaghiya estate. What is the position of the defendants in this case from the legal point of view in this light ?

34. Basra court plea

tion. Ahmad Pasha on behalf of himself and his wife Ayesha, caused their agent Abdul Wahab Pasha (otherwise known as al Qirtas) to register one-sixth of Bashiyah. The British Consul was witness. Shafiqah, being a woman, could not come before the Consul and her documents therefore remained uncertified. Transfer dealings were postponed until she and her minor sons Hasan and Abdul Bagi had died. Rafiyah, her daughter, roused by mischief-mongers, has now filed a suit. This property was bought by deed still in our possession and has been occupied by us and our forbears for twenty-four years. Now the Iraq government has begun to move against us. We ask HM Government to protect our interests."

As though the matter was not complicated enough, the Political Agent now revealed that part of the land involved was not registered in tapu because of British insistence that it be secretly registered at the British Consulate in Basra.

That was not all. In listing the defendants as the descendants of the progeny of Shaikh Mubarak and Shaikh Khazal, the question arose as to who was entitled to inheritance. Following Shaikh Mubarak's accession and the murder of his half brothers, several members of his family sought refuge in Iraq. Were their issue to be deemed part owners of the estates? Excluding them, the living descendants of Mubarak the Great totalled 120 according to the Political Agent; traceable descendants of the Shaikh of Muhammerah were fewer in number but mostly resident in Persia. There was another recurring problem. How could a writ be served on Shaikh Ahmad, the principal defendant, as sovereign head of state? In the first case, papers were served on his agent, but the ruler could easily withdraw his power of attorney.

Fortunately for the judicial experts and politicians who had to unravel the tangle, the law was slow to perform. As the Abdullah Zuhair case dragged from one adjournment to another, Iraq used every conceivable opportunity to rub salt into this and other sores. It announced that the "Ishtilak" or land tax, which so far had amounted to about £1,000 a year based on the properties worked on behalf of the Shaikh, would be replaced by an export tax on all products consumed by the Sabah family and sold elsewhere. Dickson commented: "A most deplorable atmosphere has been caused throughout Kuwait by the Bashiyah incident. There is a growing feeling that the Iraqis are fomenting these and other troubles. Yet Kuwait has gone out of its way to live in friendship with its northern neighbour since an Arab government was set up there."

The argument over the Faddaghiyah gardens was given some clarity by a letter written to Shaikh Ahmad by Abdul Aziz al Salim al Badr, who had acted for Shaikhs Mubarak, Jabir and Salim before him. Written on 15 June 1932, it said:

"In the year 1908 Ahmad Pasha bin Qasim al Zuhair, a prominent landowner in Basra, came to Kuwait and, with power of attorney from his wife Ayesha, then resident in Constantinople, sold the land named Faddaghiyah of 400 jaribs to Shaikh Mubarak. The transaction took place in Kuwait and I was ordered by Shaikh Mubarak to make registration and obtain tapu deeds."

Nevertheless, Judge Alexander did not take a very sanguine view of the outcome. Looking back at Bashiyah he commented: "The Shaikh, not being a Turkish subject, could not register it legally." And he added: "If the Faddaghiyah case comes to court he will lose it."

Anticipating that contingency, the British government began to assess the likely compensation. The Foreign Secretary, Sir John Simon, approached the Secretary for India, Sir Samuel Hoare, to see if his department would contribute. He received a very definite *No* in response. The exchequer would have to find the money. The Political Resident was told to ask the Shaikh to keep "a discreet record" of any payments made.

It was not until 1936 that the case brought by Abdullah al Zuhair finally achieved an airing in court, by which time the action was joined by Najiba and Aida, daughters of Ahmad Pasha and Ayesha. A year later, the Basra court served papers on Shaikh Ahmad's solicitor, in the absence of his agent. The British government thought this highly irregular and proposed to advise the Shaikh to declare that the man in question, Abdul Jelil Partu, had no authority to act on his behalf nor, in particular, to accept service of the summons. Not knowing the British government's intention, however, Shaikh Ahmad dismissed Abdul Jelil, thus confirming in the view of the court the validity of service. There followed a procession of court hearings, adjournments, charges and counter-charges which must be very nearly unique in legal history. The British judge presiding over the civil court, Mr H. I. Lloyd, let it be known that his fellow Iraqi judges and all the lawyers briefed in the case were corrupt and were agreeing among themselves how the proceedings would go and how the spoils would be divided. Britain wanted an Egyptian lawyer to represent the Shaikh as a non-pleading adviser, since Iraq insisted that a foreigner could not plead in its courts. Eventually it was decided to ask Salman Shaikh Daud, a distinguished Iraqi advocate, to attend the hearings. On 5 January 1938 when it was believed that Daud was on his way, the court at Basra convened suddenly and promptly adjourned *sine die*.

Britain, in preparing its side of the story, had decided that under the terms of the 1914 bequest, two-thirds of the gardens in dispute belonged to all the heirs of Shaikh Mubarak, and one-third to all male heirs only, a different view from the one put forward by Dickson. The pledge was said to cover all the lands originally given to Shaikh Mubarak unless they passed from his heirs voluntarily. To heap further confusion on the affair, the Iraqi lawyer Daud decided that he could not handle the case. Britain's ambassador in Cairo was then asked to seek out an experienced lawyer and Maitre Tewfik Doss Pasha was appointed at a fee fixed by the treasury at £400 plus daily refreshers of £40. Doss Pasha demanded £2,000 and so the search for a legal consultant started all over again. The Cairo ambassador finally secured the services of Maitre Ahmad Rushdi Bey, who left for Kuwait on 14 March 1938.

The defence, prepared by Rushdi Bey, wanted to exhibit the letter to Shaikh Mubarak of 3 November 1914, and to plead that Britain, being in effective control of Iraq up until 1922 was fully entitled to make the gift, and that all subsequent treaties and agreements between Britain and Iraq, especially those of 1926 and 1930, had bound Iraq to honour this and other international commitments entered into by Britain. The British government insisted that only part of the original letter from the Political Resident should be produced, however, since it also made reference to secret wartime agreements which gave rise to the tax concessions. The matter eventually came before the Basra court on 23 March 1938 and adjourned immediately after the defence submission. After several more resumptions and adjournments, the case was concluded in favour of the plaintiffs on 17 May and judgment delivered on 15 June 1938. It was that the clients of Partu should cease to disturb the heirs of Ayesha al Zuhair in certain plots in Faddaghiyah, should pay the costs of the hearing, and the plots be registered in the name of the heirs of Ayesha al Zuhair; existing deeds Nos. 13 to 18 dated August 1919 (presumably registered after the British entry of 1915) should be cancelled. Counsel for Shaikh Ahmad then asked that the case should be referred to the Land Claims Settlement Committee set up in Fao. This plea succeeded and Britain agreed to abide by the Land Settlement decision.

The land settlement law itself was highly complex. The original edict of 1932, passed while the first al Zuhair case was before the Basra court, protected the freeholder so long as ownership could be established. It did not refer to nationality. A new law, passed in 1938 during the later court proceedings, provided that possession could be established by persons who had planted or worked the land for not less than ten years prior to a land settlement decision, in which case it was registered as miri or crown land and the tenant became, in effect, the leaseholder. Again, in 1941, a nationality clause was enacted which deprived foreigners of the right of registration. Through the intervention of Britain, this provision was waived in the case of the ruler of Kuwait but not of other members of the Sabah family.

In the course of the appeal to the Land Settlement Committee, through the Court of Cassation, it was established that the Faddaghiyah gardens were properly registered in the name of Shaikh Jabir bin Abdullah al Sabah, who died in 1859, one plot as freehold, the others as crown land granted in tapu. In 1943, the case was resolved in favour of Shaikh Ahmad. The decision was notified in the Iraq Government Gazette of February 1944. Then came objections and appeals from tenants, nearly 3,000 in all. Meanwhile, Kuwait and its ruler were deprived of produce and income from the gardens. Their losses were estimated at £50,000. Tenants obtained possession of accretions of the property, in the direction of the Shatt al Arab, to which the owner of the land would normally have been entitled. The decision of 1943 was modified by the Appellate Court in 1949. This time it was held that one plot belonged to the Sabah family in its entirety but of the other six plots half belonged to the tenants while Shaikh Ahmad was entitled to one-twenty-sixth part of the remainder. Although there were many entitled descendants, only the ruler was deemed by the court to have a right of registration. This decision was rejected by a majority view in the Court of Cassation, with the British presiding judge dissenting, in 1950, when it was held that ownership of the one plot should not be vested in the ruler.

The ruler and the Sabah were effectively deprived of 180 out of 216 shares in the Bashiyah property, which was considered part of the Faddaghiyah estate by the court. Britain asked Shaikh Ahmad to submit a claim for compensation and a sum of £8,400 was agreed, the price paid for the repurchase of shares following the court proceedings. Earlier, in 1940, the British government paid Shaikh Ahmad £30,694 in restitution of losses suffered by several heirs of Shaikh Mubarak through Ishtilak tax imposed up to that time. In return, Britain was absolved from its earlier guarantees in connection with the Faddaghiyah gardens.

The rash of litigation was not over yet. No sooner had the Faddaghiyah case ended than another action started up over the Fao gardens. This reached the stage of appeal after another ten years, and disappeared into limbo in 1953. Shaikh Abdullah, Kuwait's new ruler, visited London for the Coronation in 1953 and used the opportunity to discuss the matter with the British government. In the previous year he had visited Baghdad where he raised the question with the Iraqi prime minister. There, as far as the record goes, the matter ended. It was a welcome demise. Britain had discharged her obligation and Kuwait, by this time, was able to look on the matter more as a question of principle than of need. The heirs of Shaikh Khazal received nothing, and Britain showed in the memoranda of the day a distinct sense of disquiet. The complications of this costly litigation often defied belief. The lawyer Partu, re-engaged by Shaikh Ahmad after his initial dismissal, was called up for military service at the time of the first appeal, but he managed to combine the duties of an Iraq army officer and solicitor to the ruler for three years. Then at a critical stage in 1942 he finally gave up his part in the proceedings when he became mayor of Basra. The unfortunate Aida, one of the beneficiaries of the court judgment in the case initiated by Abdullah al Zuhair, had spent her expected fortune in Constantinople and was immediately pounced upon by her creditors when the appeal decision became known. As it happened, she died in the interval of the court hearing before an attachment order for 10,000 dinars could be taken out against her. Early in the proceedings the authorities in Baghdad made a tax claim on Shaikh Abdul Chassib for £19,480, but in the legal hurly-burly that followed it seems to have been lost sight of. And one appeal was lost by the defendants because Partu used the wrong tense in the wording of a surety. It was altogether a remarkable legal case.

Internal pressures

Other issues of this period tried the patience of Shaikh Ahmad and sometimes disturbed the outwardly placid life of Kuwait.

The Iraqi press, which had for years pounced on the slightest excuse to attack its smaller neighbour, built up a steady campaign in the late 1930s designed to stir up disaffection in Kuwait, and to give as much substance as it could to the suggestion that the country's economic difficulties arising from the fourteen-year Saudi blockade were the result of maladministration by the ruler.

In addition, German propaganda emanating from Baghdad had brought about the establishment in Kuwait of a young men's revolutionary movement known as Al Shabiba. Partly as a result of the Iraqi press campaign, partly through the Shabiba party, and partly from the feeling among leading businessmen and other members of the community that the Shaikh made no provision for consulting them on some vital matters, there was a mood of unrest in Kuwait by early 1938. A group of foreign citizens began to call for action against the Shaikh. The crucial oil strike had been made in February of that year, and it was no coincidence that in March the agitation reached a dangerous pitch.

Italy at this time was playing a suitor's role with the nations of the Middle East and Arabia by way of short-wave broadcast transmissions. Early in the new year somebody close to the ruler, though exactly who was never determined, arranged for Bari radio to put out a broadcast beamed to all the Arabian countries extolling the virtues of Kuwait. It gave a fulsome account of the country and its ruler, who was described by the Italian broadcaster as a leader as just as Abu Bakr, as courageous as Omar, standing fearlessly in defence of Arabs everywhere, and steadfastly opposed to British machinations in Palestine. The Political Resident, Sir Trenchard Fowle, said it was "like a description of Heaven and the Angels". Perhaps so, but it is hard to understand why this particular event excited the British government so much. Admittedly, somebody in Kuwait must have furnished the Italians with material for the broadcast, but more extravagant things have been said of football teams on British radio than were said of Kuwait that day. At any rate, the Political Agent was instructed to convey the deep displeasure of HM Government and to express the hope that if Shaikh Ahmad wished to publicise his country in future he would use the BBC. The Shaikh replied that he could see no harm in people singing the praises of his country.

That incident is worth recalling in view of the later anxiety of the Political Agent and Sir Trenchard Fowle to dilute the ruler's authority.

By March 1938, agitation, fanned by broadcasts and secretly distributed leaflets, had become persistent. Some of the ringleaders beat a retreat to the safety of Baghdad, but one of them was caught and publicly flogged. Such punishment was not uncommon in Arabia, but in the prevailing mood it was made the pretext for a local outcry against autocratic rule, and the powerful group of citizens who voiced their disapproval had the added satisfaction that the flogging had provoked a protest from the British government. The whole incident brought a renewal of demands for some form of representative assembly or elected council, a subject which had been discussed at fairly regular intervals among the merchant community ever since the original attempt at forming a council during the first year of Shaikh Ahmad's reign. The new movement was given most of its impetus by the determination of powerful merchant families to have a say in matters of government.

On hearing of the disaffection in Kuwait, the British government, through the Political Resident, encouraged Shaikh Ahmad to meet some of the demands for financial reforms, and to develop educational and hospital facilities. A Foreign Office missive to the Political Resident pointed out primly that "agitation is best dealt with by sympathetic guidance into useful channels of activity".

It was perhaps unfortunate that at this period relations between the ruler and the Political Agent, Captain de Gaury, were less than cordial. The Political Agent felt that the time was past for one-man rule in Kuwait, and was pressing the Shaikh to set up a consultative body. Shaikh Ahmad, proudly on the defensive, refused to be told

35. At Bahrah, 1936: de Gaury, Dickson, Holmes and Shaikh Ahmad

how to run his country. In the end, however, the Shaikh gave in to mounting pressure, and on 29 June 1938 a Majlis-at-Tashrii was formed. On 9 July a law governing the powers of this newly formed Administrative Council was signed by Shaikh Ahmad. It began: "The people are the source of power, as represented by the council, their elected representatives." There followed a series of laws relating to the budget, justice, public security, education, health and national emergencies. It was, in its way and in the context of Arabian political tradition, a progressive document. Shaikh Ahmad, a conservative ruler who believed more in traditional Arab ways of government than in imported ideology, nevertheless adjusted himself to the new situation, and after a few months asked the council to take over the financial affairs of the state, though these had always been his personal prerogative.

In September, a constitution was prepared by the council describing the Kuwait flag and nationality and underlining the contract by which Britain assumed responsibility for the country's protection.

The new popular movement had some strange consequences. Iraq, unable to claim any longer that it was opposing a repressive regime in Kuwait, decided that the Legislative Council was an imperialist trick on the part of Britain, and turned its attention to the smuggling issue. A puzzled Foreign Office, underestimating Shaikh Ahmad's capacity for empirical decision, found it hard to believe that he was allowing the Council to administer the state's finances. However, these were not exactly the nuptials of democracy. The Shaikh retained very definite powers and opinions. In August the council questioned the activities of his private secretary and a battle developed which resulted in this Syrian member of the Shaikh's entourage, Izzat Jaafar, hurrying off to Iraq. For a few days Captain de Gaury assumed that the young Syrian was the sole cause of the trouble and that his disappearance resolved the matter. It transpired, however, that the aged Mulla Saleh – the Shaikh's closest adviser and confidant – was also the subject of animosity on the part of one of the contending factions. In the end, it was agreed that the old secretary should go on indefinite leave. A long tug-of-war ensured and the ruler's patience finally snapped. He dissolved the council on 21 December. Nine days later a new council was formed with twenty members, twelve of whom had served on the old council, under the presidency of Shaikh Abdullah al Salim. Again, feeling was aroused by Iraqi agitation, this time involving a tailor, Sayid Fakhri, who was said by de Gaury to be insane. He had telegraphed to his home in Iraq saying that his life was in danger, to the delight of the Iraqi press. In the troubled circumstances of the time, the second Administrative Council never got off the ground. In any case, its members could not agree to the new draft constitution presented by the Shaikh. In its place, the ruler set up an Advisory Council consisting of four members of the Sabah and nine notables, again under the chairmanship of Shaikh Abdullah al Salim. It came into being on 14 March 1939. A few days earlier, on 10 March, another clamour was caused by an impassioned public speech denigrating Shaikh Ahmad. The culprit was promptly and predictably arrested. While on his way to the gaol, two other men tried to release him. A battle broke out and the Shaikh's bedu bodyguard, angered by the propaganda of past weeks, attacked the prisoner and his supporters. The former was promptly tried and executed.

Thus ended the first and only event involving serious public disorder or bloodshed in the 200 years of Kuwait's history as a shaikhdom. In October 1939 a new Political Agent, Major Arnold C. Galloway, was appointed, and a new Resident appeared in the Gulf, Lt Col C. G. Prior.

The reformists in Kuwait, though powerful, did not represent a majority opinion. It is worth remembering that the bedu, and they included the ruler's men-at-arms and his personal bodyguard, were solidly behind Shaikh Ahmad. The Shia Persians in the town, and there were several thousands of them, also disliked the council.

At the time of the dissolution of the council at the end of 1938 the balance of power shifted in favour of Shaikh Ahmad. Britain seemed pleased at the outcome. But this was by no means the end of Kuwait's troubled times. Water from the Shatt all Arab, arms smuggling, and geographical disputes which had been hoary issues for more than half a century, came to the fore in a manner that was to find echoes in the politics of the 1960s.

Iraq claimed that arms, tea, and other goods to the tune of a quarter of a million pounds a year were being smuggled in from Kuwait, though it was likely that much of the contraband came from Saudi Arabia. Because of this smuggling Iraq protested a right to control Kuwait's customs, an issue that had been raised even in the days of Shaikh Mubarak, on the ground that Kuwait was part of the wilayat of Basra (an Ottoman province) prior to 1914. It was claimed that the country should be incorporated as an autonomous shaikhdom within Iraq. The Iraqis also suggested that a pipeline should be laid between the Shatt al Arab and Kuwait. Perfidy was not hard to detect, and both Britain and Kuwait rejected the overture out of hand.

6. Talking of Oil

Oil gave a new aspect to Kuwait, and to the entire Arabian peninsula. For hundreds of years Kuwait looked into the desert, to the Najd and beyond to Mecca, for spiritual inspiration; and seaward, across the Gulf, for its trade and contact with the outside world. Then, suddenly, it embraced an alien technology.

It took on its new dimension with compulsive enthusiasm, using its enormous wealth with a liberality that few nations had shown before and few are likely to show in the future; gathering technical skills and administrative capability as fast as its growing prosperity and limited manpower resources would allow.

But if the advent of oil in Kuwait was greeted with a flurry of activity, its coming was exceedingly slow.

References to black, oily substances were common among travellers and in the literature of Arabia. The uses of bitumen throughout recorded history were manifold. In 1869, ten years exactly after the first American "strike", oil was discovered in Egypt. Nearly forty years later, in 1908, the prodigal well at Masjid-i-Sulaiman, the Temple of Solomon, in Persia began to flow. It was not for another thirty years, however, that an even more fertile field, probably the most abundant of all, was discovered in Kuwait. The reasons for the delay are bound up in the processes of political and commercial bargaining which in the Western world, where the necessary technology lay, can be long drawn out. Indeed, it took two world wars to provide America and Europe with the incentive to develop resources which, in many places, literally oozed out of the ground. The first of those wars brought some very powerful men of politics and commerce into the picture. But the scene was set in the last two decades of the nineteenth century.

So far as our chapter of the long and well-documented story of oil is concerned, we must go back to the man who gave the initial push to the search in the Middle East, William Knox D'Arcy. His father, an English solicitor, took his family to Australia in the 1860s. William qualified and joined his father's legal practice where he showed a considerable aptitude for recognising a business opportunity. In 1882, when he was thirty-two years of age, he was consulted by the brothers Edwin and Thomas Morgan who at the time were prospecting for silver. As it happened they found gold and D'Arcy, who provided – or at any rate found – the financial backing needed to mine the metal, became a substantial shareholder in the business. By 1890 or thereabouts he had made a million pounds and was looking out for ways of increasing that sum. He returned to Britain where, in a social whirlpool that contained every kind of commercial opportunist, he met Sir Henry Drummond Wolff, sometime British Ambassador to Persia. Some years earlier, a French geologist, Jasques de Morgan, working with an archaeological expedition in Persia, noticed some very evident oil seepages. He recorded his find and filed it away. It was not until ten years later, in 1900, that he decided to look for a financial backer to enable him to investigate the matter more closely. News of de Morgan's search reached Sir Henry who in turn talked to D'Arcy. There were several chance meetings between times, but that was the essential chain of contact which led to an exclusive concession for a British company to seek oil in Persia. Concession agreements are notoriously difficult to negotiate, however, and even when the sordid details of payment and profit distribution have been worked out there remains the matter of finding the oil, as many subsequent prospectors were to discover.

Persia was no exception. Three men undertook the bargaining. They were Alfred Marriott, cousin of D'Arcy's secretary G. W. Marriott; Edouard Cotte, a Frenchman with a good knowledge of Persia and its politics; and General Antoine Kitabji, ex-director of the Persian Customs Service. Their talks with the Shah's Grand Vizier were at first smooth and trouble free. Britain's Minister in Tehran, Sir Arthur Hardinge, supported the negotiators by way of demonstrating the concern and interest of HM Government. A contract was no sooner ready for signature than the Russian Imperial government protested. Sir Arthur Hardinge in turn protested. Mr Marriott tried an undiplomatic but well-conceived way of ending the deadlock. He offered an impoverished Persian government £10,000 of D'Arcy's money in return for an immediate signature. Many tales have been told of the ingenious methods employed to hoodwink the Russians, whose concern at the prospect of British-backed developments along its southern boundary at a time when it was pursuing similar "rights" in Persia, had earlier contributed to the annulment of a concession granted to Baron de Reuter, founder of the famous British newsagency. Whatever the methods used, the fact remains that an agreement was signed on 28 May 1901. Its eighteen clauses were written in French and Persian and the document was signed by Marriott for William Knox D'Arcy. The Persian government was to receive further payments of cash and shares as well as 16 per cent of net profits. D'Arcy contracted to form an exploitation company within two years. All the negotiators, including Sir Henry Wolff, were suitably rewarded. De Morgan was to receive a share of Cotte's bounty. The company was formed in 1903. Thus, pioneers began to search in remote and usually lawless regions of Persia for commercial quantities of oil.

Perhaps the most important of these pioneers was G. B. Reynolds, a driller of some experience, a man of enormous persuasiveness and organising ability and no little obstinacy. He is not directly relevant to our story, except in the vital fact that without his tenacity – which led him even to disobey the instructions of his employers – D'Arcy's oil venture would have been doomed to an early demise, and they would have been in no position some thirty years later to help find and exploit the oil deposits of Kuwait.

Sir Arnold Wilson, who as Lieutenant Wilson met Reynolds when he took a detachment of Indian soldiers to Ahwaz in Persia, wrote: "He was dignified in negotiation, quick in action, and completely singleminded in his determination to find oil. . . . He was a great Englishman."

While Reynolds drilled first one dry well, then another, D'Arcy's fortune diminished. The financier, back in London, neared the point of no return. He had spent more than £250,000 in cash and used up most of his Australian mining shares as security against loans. He looked around for a purchaser of his share of the Persian oil rights. At this stage, Lord Fisher, First Sea Lord, entered the fray. He was concerned about the availability of fuel for his warships, and induced the Burmah Oil Company, based in Scotland, to form a syndicate with D'Arcy. As further drilling revealed more dry holes and money continued to dissolve in Persian air, the enthusiasm of that company waned noticeably. By 1908, funds were exhausted. It has been said that after vain drilling attempts at Chiah Surkh in north-west Persia, and Shardin in the south-west, the syndicate decided to call it a day and sent Reynolds a message from London to that effect. Whether such a message was in fact sent, or indeed received, will never be known for certain, though Sir Arnold Wilson affirmed it with the acid comment: "Cannot government be moved to prevent these faint-hearted merchants, masquerading in top hats as pioneers of Empire, from losing what may be a great asset?"

Whatever he may have been told to do, Reynolds took matters into his own hands and transferred the rig from Shardin to Maidan-i-Naftun, near the Temple of Solomon, eighty miles away. He had wanted to work this site from the beginning and it was here that de Morgan had seen manifest signs of oil – but he had been told by London to go on with the Shardin exploration.

On 26 May 1908, he struck oil in commercial quantity at 1,200 feet. Reynolds was vindicated, the Burmah Oil Company and D'Arcy were saved from virtual bankruptcy, the Royal Navy's bunker fuel was assured. The economies of Persia and Britain were about to be transformed. A year later, in April 1909, the Anglo-Persian Oil Company came into being with a capital of £2 million. The scenes in London and at the Bank of Scotland offices in Glasgow were chaotic as the public fought for the 400,000 preference shares on offer. The issue was fifteen times over-subscribed.

Meanwhile, the new company began to marshal its resources of manpower and equipment in Persia. Some of the senior personnel sent to Masjid-i-Sulaiman, or MIS as it became to oilmen the world over, were later to apply their enormous experience, much of it gained from the indefatigable George Bernard Reynolds, to the exploration and planning of the Kuwait fields. One in particular was the brilliant Scottish engineer James A. Jameson, who went to Persia as assistant to Charles Ritchie, the chief engineer.

The Persian discovery did not mark the end of Admiral Lord Fisher's interest in the matter of oil supplies. The Admiralty committe on oil over which he presided in 1904 was resurrected in 1912 by a new First Lord, Winston Churchill, who shared Fisher's passion for ensuring that the country's warships should not go short of fuel in peace or war. Lord Fisher again found himself at the head of the

Royal Commission on Oil for the Navy, while a committee of experts led by Admiral Sir Ernest Slade and Professor John Cadman from Birmingham University sailed for Persia and the Gulf in order to report on future prospects.

One of Admiral Slade's engagements of that tour was a meeting with Shaikh Mubarak of Kuwait. On 23 January 1899, the ruler of Kuwait had signed an agreement with Britain which entitled his country to Britain's protection in return for an assurance that neither he nor his heirs would "cede, sell, lease or mortgage, or give for occupation or for any other purpose a portion of his territory to the government of any other power without the previous consent of Her Majesty's Government".

In 1913, in anticipation of Sir Ernest Slade's visit, Shaikh Mubarak wrote to the Political Representative in the Gulf, Sir Percy Cox: "We are agreeable to everything which you regard as advantageous . . . we will associate with the Admiral one of our sons to be in his service, to show the place of bitumen in Burgan and elsewhere and if in his view there seems hope of obtaining oil therefrom we shall never give a concession to anyone except a person appointed by the British government."

The special relationship between Kuwait and Britain was cemented a year later when, with the outbreak of hostilities, the Turkish Empire allied itself with the Germans, while Kuwait's ruler honoured his agreement with Britain by opposing the Ottoman occupation.

In May 1914, the British government was taking another step in the direction of the Middle East. Churchill, to whom the Royal Navy was perhaps more sacrosanct than any of the other causes he was to champion in the course of a long and hard-fought political career, decided to protect the supply lines of that institution by means of government investment in the Anglo-Persian Oil Company. He presented a bill to Parliament which would give the nation a 51 per cent shareholding at a cost of just over £2 million. The bill was passed with a majority of 236 votes and received the Royal Assent six days before the declaration of war. A further £3 million was invested in 1919. Within fifty years that capital outlay was to appreciate some 120 times, to a value of more than £600 million. Kuwait was destined to play an important part in that appreciation and in affairs which would soon enmesh much of the Arabian peninsula, the USA, Britain and Europe, and many of the greatest commercial enterprises in the Western world, in a relationship that to say the least of it was delicately poised.

Meanwhile, discoveries were being made across the Atlantic Ocean which proved to have a bearing on the story of Kuwait's oil exploitation. Again, like the discoveries in Persia, they followed an almost endless series of disappointments and frustrations.

On 10 January 1901, Anthony F. Lucas was in charge of a drilling operation at a place called Spindletop in Beaumont, East Texas. They had reached 1,020 feet when, with little hope or enthusiasm, they withdrew the drill string in order to replace a worn bit. They had hardly restarted the drilling operation when an unexpected roar heralded a gush of oil that rose 200 feet or more into the air, covering the surrounding countryside with a thick black lava. The well was eventually brought under control and the oil industry of Texas was born. As with all exploration, large sums of money were needed to buy expensive equipment and maintain the ever optimistic spirits of those who used it. Lucas had enlisted the help of two Pennsylvanian oil prospectors, James M. Guffey and John H. Galey. They borrowed 300,000 US dollars from the banking house of T. Mellon and Sons, one of whose directors, Andrew W. Mellon, was to become US Secretary to the Treasury and Ambassador to London.

The inevitable shifts of equity and power came into effect, with the Mellons taking over the assets of the original group, while Guffey purchased the interests of Galey and Lucas. The J. M. Guffey Petroleum Company was formed.

The next step was to create a refining company. Guffey and the Mellon brothers, A. W. and R. B., wanted to form the Texas Oil Company, but the name was already registered. With the Gulf of Mexico as the nearest geographical point of any significance, they decided to call it the Gulf Refining Company of Texas. It was registered on 10 November 1901. In that year crude oil production reached 670,000 barrels. Construction of a refinery at Port Arthur began the same year. In 1907 the company was again reconstituted, with the incorporation of the Gulf Oil Corporation in New Jersey. A. W. Mellon became its president. By 1908, assets were valued at 18 million dollars. From then on, its expansion in the United States was rapid. By 1921, its assets had increased more than tenfold to 202 million dollars.

With the conclusion of the first world war in 1918, most of the participants were exhausted, and many of those who had maintained a neutral position or by good fortune had avoided direct involvement were no better off. Lord Curzon is quoted by Henry Longhurst, BP's biographer, as saying: "The allies floated to victory on a sea of oil."

The country that supplied much of that oil and the nations that used it to gain a military victory, profited little. Persia remained poor and hopelessly corrupt. The Ottoman Empire lay in ruins. Russia underwent a revolution. Britain, despite her extant empire, and the United States, despite her vast wealth, moved rapidly towards economic collapse. Shortly, the quest became one for new uses for existing oil supplies, rather than for new sources of supply. Nonetheless, far-seeing men in government and industry, as well as those with a long-term eye to the main chance, still looked to oil as a source of wealth and power which would eventually help to put the world on its feet. Britain certainly had no intention of resigning its interest in the Persian fields, though there is evidence to suggest that it was less than enthusiastic about extending those interests. In 1921 a Persian army officer Brigadier Reza Khan seized Tehran and usurped the last of the shahs of the Qajar (or Kadjar) dynasty, whose kings had occupied the throne of Persia since Aga Muhammed Khan in 1795. He was proclaimed Shah of a new dynasty, the Pahlevis, in 1925 and became known as Reza Shah. A bristling new broom swept through Persia and Anglo-Persian affairs. Local rivalries were subdued, familiar figures among the oil community disappeared, new powers came into prominence. In the late 1920s, Britain – along with the rest of the world – suffered a trade depression in which oil sales declined dramatically. By 1931, Anglo-Persian's dividend had sunk to 5 per cent. Two years before it was 20 per cent. Persia's income was affected by this depreciation, and talks which began in 1929 with the aim of revising the original D'Arcy agreement, giving the Persians 16 per cent of net profits, were suddenly called off. The new Shah annulled the concession in 1932.

The dispute was referred to the League of Nations and Dr Benes, the great Czechoslovakian statesman, himself to make desperate appeals to international arbitration a few years later, was asked to report to the Council of the League. Dr Benes suggested that the two sides try to settle the matter themselves and in 1933 Sir John Cadman, the company's chairman, and Mr W. Fraser, its deputy chairman, met the Persian side in Tehran. They failed to agree in committee, but as the British delegates called to say goodbye, the Shah realised that intransigence was not going to pay the rent. He took personal charge of the bargaining and within a few days, a new agreement was drawn up. It scrapped the simple percentage-of-profits arrangement and gave Persia a royalty of 4 shillings a ton on exported oil, together with additional payments based on tonnage in lieu of Persian taxes. Also, Persia was to receive 20 per cent of dividends on ordinary shares over an agreed minimum figure, the company guaranteeing an annual payment of not less than £1,050,000. The original concession, which was to terminate in 1961, was extended to 1993. Mr Fraser, later the first Lord Strathalmond and an important figure in the story of Kuwait's oil concession, regarded the new agreement as one of the fairest ever concluded. It was to form a model for other pertinent negotiations.

While Persia and the Anglo-Persian Oil Company were ironing out their immediate problems, much was going on elsewhere that was to affect the future of both, and indeed of the rest of the world.

The thin red line

Perhaps more significant to the minds of oil men than the discovery at Masjid-i-Sulaiman in 1908, or the constitution of the Anglo-Persian Oil Company out of a hotch-potch of D'Arcy men and the Burmah board in 1909, was the laying of a pipeline from the oilfields to Abadan on the Gulf in 1911. Talk to anyone in the industry, be he labourer or technician or managing director, and it will not be long before he mentions Abadan. Not Persia or a particular oilfield. Abadan. To the oil communities of the Old World it is, in a way, what Drakesville is to those of the New World. A focal point. A womb.

Britain was almost parochially Abadan-conscious when it came to taking stock of oil resources and potentialities. But no such consideration entered into American thinking. North America had a thirst for oil unmatched by the rest of the world put together. By the 1920s the "open door" approach was very much part of US policy.

The agreement of 1901 had specifically excluded the northern territories of Persia from the D'Arcy concession, but as is the way in oil matters – which usually transcend purely commercial considerations – possession became nine points of the law. Russia, after the revolution, renounced Tsarist claims in those regions on the understanding that no future concessions were granted to third powers. But it did not take the Persian government long to realise that it had a considerable sleeping asset on its hands. It opened negotiations with two American companies, Standard Oil of New Jersey and Sinclair Consolidated Oil Company. In 1923 an agreement was signed between Persia and Sinclair. The latter had an exclusive agency for the sale of

Russian products abroad and were, it is said, in a position to pacify the Soviets. However, a public incident in which an American diplomat was done to death by an angry crowd, made the firm's position untenable and the Sinclair concession was abandoned in 1925.

The Americans, aggrieved by the refusal of this door to open, looked elsewhere. Anglo-Persian maintained its right to exclusivity in the country of its origin, citing the Russo-Persian agreement to that end, but the British company kept an eye on developments in and around the Arabian peninsula. The most important of these was the setting up of the Turkish Petroleum Company to exploit the resources of Mesopotamia – or Iraq as it was to become.

Following the first world war this Ottoman region was under British influence, and at the San Remo conference of 1920, Mesopotamia was mandated to Britain. Part of that agreement was a division of oil interests in the area between Britain and France. The British signatory was Sir John Cadman, at that time head of the Petroleum Department on the Board of Trade. Sir Percy Cox, perhaps the most professional and certainly the most ubiquitous of all Britain's eastern emissaries, arranged for the Amir Faisal, the son of King Hussain who was T. E. Lawrence's ally in his Arabian campaigns, to occupy the throne. The country officially became Iraq, and its new king was crowned on 23 August 1921. Before the British mandate, oil had attracted a good many sightseers to the hills and plains through which the Tigris and Euphrates flowed. New generations recalled biblical references to the black bituminous deposits of the area, though none was yet aware of even earlier references in the Sumerian tablets of Ur and Nippur which told of Utu-nipishtim's salvation from the Great Flood, in an ark loaded with creatures and protected from the water by bitumen.

The interested parties in Iraq had divided themselves into five main groups prior to the outbreak of the 1914–18 war. They were the Deutsche Bank, the D'Arcy consortium, the Royal Dutch-Shell Company, the American Chester Group, and African and Eastern Concessions, later the Turkish Petroleum Company.

The German company was primarily concerned with the long-proposed Berlin–Baghdad railway which, in fact, was intended to terminate on the Gulf, in or close to Kuwait. Mr Calouste Gulbenkian had shown an interest in Mesopotamian oil. Admiral Colby M. Chester had obtained railway construction and mineral concessions from the Ottoman Turks in 1910, though they were never ratified. The French, too, had an interest in the matter since they controlled the only possible outlets on the Mediterranean, by virtue of their mandate over Syria and the Lebanon. But it was the Armenian Gulbenkian who called the tune in this oil dance. He had helped form the National Bank of Turkey with British capital, together with Sir Ernest Cassel. In 1912, the African and Eastern Concession, owned by Cassel, became the property of the National Bank and the Deutsche Bank, and its name was changed to Turkish Petroleum Company. Shareholding was divided between Deutsche Bank 25 per cent, Royal Dutch Shell 25 per cent, C. S. Gulbenkian 15 per cent, Sir Ernest Cassel and National Bank of Turkey 35 per cent.

A year later, Sir Ernest Cassel placed his shareholding at the disposal of the British government, to the consternation of Henry Deterding, chairman of Shell. The Dutch and British governments were now in conflict and in order to keep the peace, Gulbenkian offered 10 per cent of his holding in Turkish Petroleum to Deterding. Thus, he became a five per cent shareholder – with which quantity he was later to be dubbed – in the new scheme of things. The National Bank of Turkey's share, which was of course acquired with British money in the first place, was taken up by the D'Arcy group and by April 1920, when the company was officially reconstituted, Anglo-Persian had become the largest shareholder with 47·5 per cent, against Shell's 22·5 per cent and Gulbenkian's five. The other 25 per cent was allocated to a French consortium. America was still out in the cold.

In 1922, Mustapha Kemal, fresh from his victory over Greece, abolished the Sultanate and with it any claims to the old Ottoman territories, with the exception, it later transpired, of Mosul. However, the inclusion of this territory in Iraq, under British mandate, was accepted internationally and Turkey was in no position to use force. By this time, oil technicians began to move in to Iraq, spurred on by the discovery by Anglo-Persian of commercial deposits at Naft Khaneh on the Iran–Iraq border. America intensified its demand for a share in Middle East oil exploitation, with the State Department reiterating in no uncertain terms its much-quoted open door policy. In October 1927, Baba No. 1 well at Baba Gurgur near Kirkuk exploded into life and the State Department advised the American oil companies, banded together in the Near East Development Corporation, to effect a compromise agreement with the Turkish Petroleum Company. Again, Gulbenkian was the central figure in the negotiations. In 1928, the company was reconstituted so that Anglo-Persian, Shell, Compagnie Francaise des Petroles and the Near East Development Corporation held 23¾ per cent of the shares apiece. Gulbenkian retained his five per cent. Within twelve months of these changes, the organisation became the Iraq Petroleum Company.

The companies at this time making up the Near East Development Corporation were Standard Oil Company (New Jersey), Socony Oil Company, Gulf Oil Corporation, Atlantic Refining Company and Mexican Petroleum Company. Sinclair and the Texas Company were originally members of the consortium but withdrew at this stage.

A concomitant of the new agreement was a pact between the consorting companies limiting their individual activities within an area marked in red on the map. Thus it became known as the Red Line Agreement, a document which was to have an important bearing on the negotiations for oil exploration rights in Kuwait.

Preparing the way

The red line embraced a large part of the Ottoman Empire, from Turkey in the north to Yemen and Hadramaut in the south, from the Red Sea in the west to the Gulf in the east. But it excluded Kuwait. In effect, the agreement formed part of the Turkish Petroleum Company reconciliation of 1914, although it was not called the Red Line Agreement until 1927 when its provisions were seen to place a severe restraint on the activities of at least one of the participants, the Gulf Oil Corporation. The United States government had all along protested vigorously about the restrictive nature of the agreement which surrounded the Iraq concessions and the post-war disposal of oil stakes. At San Remo, Britain had given Germany's share in TPC to

France. In return, France relinquished her rights in Mosul granted by the Sykes-Picot wartime agreement. At the same time, France allowed TPC pipeline transit facilities through its mandated Syrian territory. When, eventually, the American companies of the Near East Development Corporation were admitted to the club, they gained a vital foothold but they committed themselves to a standstill. Anyone who wanted to explore farther afield within the red line could be prevented from doing so by the others, even though they had no ambitions of their own in that particular territory. It was a mutually-agreed dog-in-the-manger approach and it was unlikely to suit everyone for very long.

In 1926, the Eastern and General Syndicate, a British company formed to obtain and sell mining concessions in the Middle East, for which Major Frank Holmes worked, offered options in Bahrain, Hasa, Kuwait and the Kuwait Neutral Zone to the Gulf Corporation.

Holmes was not to be put off by arbitary red lines on maps. In November 1927 he entered into a contract with Gulf Oil whereby all the rights of Eastern and General Syndicate were acquired by a subsidiary of the latter, Eastern Gulf Oil Company.

At about this time, the Standard Oil Company of California was looking around for oversea sources of supply since its home fields were becoming depleted and its foreign exploration ventures had all proved costly failures. The Gulf Company approached Standard Oil in December 1928 with the suggestion that the latter should take over certain leaseholds and obligations contracted by the Eastern Gulf Company in the 1927 deal with Holmes, thus freeing itself from the stranglehold of the red line agreement to which its recent entry into the Turkish Petroleum Company had made it subject. The main leasehold applied to Bahrain. Standard Oil wisely took over the option. Other Gulf options in Saudi Arabia, Hasa and the Kuwait Neutral Zone were assigned back to the Eastern and General Syndicate in 1932.

In 1928, Col J. C. More, the Political Agent in Kuwait, had written to Major Holmes advising him of a clause which HM Government desired to have included in any agreement with the Shaikh of Kuwait. It read:

"The Company shall at all times be and remain a British company registered in Great Britain or a British Colony, and having its principal place of business within His Majesty's Dominions, the Chairman and Managing Director (if any) and a majority of the other Directors of which shall at all times be British subjects, and neither the company nor the premises, liberties, powers or privileges hereby granted and devised, nor any land occupied for any of the purposes of this lease, shall at any time be or become directly or indirectly controlled or managed by a foreigner or foreigners of any foreign corporation or corporations, and the local General Manager of the company, and as large a percentage of the local staff employed by them as circumstances may permit, shall at all times be British subjects or subjects of the Shaikh.

"In this clause the expression 'foreigner' means any person who is neither a British subject nor a subject of the Shailh, and the expression 'foreign corporation' means any corporation other than a corporation established under and subject to the laws of some part of His Majesty's Dominions and having its principal place of business in those Dominions."

Meanwhile, Holmes was encouraged by the Gulf Company to make efforts on its behalf to secure a concession in Kuwait, notwithstanding the fact that the British government continued to stress the inviolability of its contract with Shaikh Mubarak and his successors whereby no oil concession would be valid which did not include the provisos set out by Col More. Britain's attitude at this time was open to the charge of ambiguity. But such an accusation should be examined in the light of prevailing world problems. Reference to British government records of the period shows that there was a justifiable concern at the prospect of over-production at a time of general economic depression. They also show an almost pathological concern for signs of Soviet Russian expansion into the oil domains of the West.

Imperial defence deliberations of the period 1928–32 made frequent reference to Soviet instructions to the Communist Parties of the Middle East to disrupt oil production. The oil policy of the Soviet Union was under constant and anxious review at this time. In 1932, a Russian oil official, Mr J. V. Volodorsk, was detained in London under the Prevention of Corruption Act. Finding "reds" under the carpet became a national pastime in a Britain beset with political and economic problems. There were several "green papers" in the early thirties relating to German arms supplies to rebel forces in the Hejaz-Najd, and to Hashemite intrigues against Ibn Saud. The British government was also in conflict with France over the Iraq concessions. It was, again, the period of Britain's rejection of the gold standard, and of the succession of the second national government under Ramsay MacDonald; not exactly a cavalier administration.

There was certainly no anxiety to increase the area of trouble abroad.

The Achnacarry agreement, often called a cartel agreement, to which Anglo-Persian was a signatory, laid down that there should be co-operation to regulate the wildly fluctuating oil markets of the world. In 1932, soon after the opening of yet another field in the Middle East with the on-flow of Bahrain oil, Sir John Cadman by now chairman of Anglo-Persian, spoke to the American Petroleum Institute. He said:

"Consumption everywhere has decreased, competition is too keen; and prices contain no margin for gain. One possible step toward rehabilitation of trade is evidently the readjustment of supply to demand and the prevention of excessive competition by allotting to each country a quota which it will undertake not to exceed."

36. A. H. T. Chisholm, chief negotiator for the Anglo-Persian Oil Company

Such were the considerations which assuredly influenced the British government and the Anglo-Persian Oil Company during the long, eventful negotiations that led to the discovery and exploitation of Kuwait's oil.

APOC chose A. H. T. Chisholm to present its case when, in mid 1932, it approached Shaikh Ahmad with a firm counter to the offers which Holmes was making on behalf of Gulf.

It has been said that Holmes and Chisholm were well matched as negotiators. If extremes of manner and appearance make a good match, then these were indeed well chosen opponents. Chisholm was half Holmes' age. But their differences transcended age. Archibald

Hugh Tennent Chisholm was not by any stretch of the imagination cut out in speech, habit or dress for the role of the entrepreneur. Educated at Westminster and Christ Church, his manner was urbane, his approach dry and witty. *Languid* is a word that has been used to portray him. Perhaps *phlegmatic* serves better. Holmes, squat, chain-smoking, often wearing a bizarre arrangement of abba over a European suit, always on the look out for a short cut to fortune, could hardly have presented a more startling contrast. It is not easy to imagine Chisholm's reaction when he put his monocle to his eye to take a first look at the adversary who, by this time, had become the "father of oil" to the Arab world. By 1932, when they came together, or at any rate to confrontation, Holmes had been instrumental in finding first water and then oil in Bahrain; and had come close to repeating the process in Kuwait. There was an element of irony in the fact that the negotiators were disposed on opposite sides of the bargaining table. Six years before, Holmes had tried without success to persuade the Anglo-Persian Oil Company to take up his Arabian concessions. The Americans had shown more sagacity.

The negotiations

The discussions which led to the first Kuwait oil concession are of such importance in the country's history and have been thus far so meagrely documented that it is as well at this point to examine in detail the complex exchanges which took place.

In 1925, when the Eastern and General Syndicate was seeking the approval of the British government in connection with the Bahrain concession, it had asked the then Colonial Secretary, Leopold Amery, whether the British would have any objection to a similar concession being sought in Kuwait. According to the syndicate, the Colonial Secretary had given them the go-ahead, subject to the incorporation of the same protective clauses as had been insisted on in the case of Bahrain. Thus, in 1925, the syndicate made approaches to Shaikh Ahmad. So did the Anglo-Persian Oil Company. Even before this date, tentative moves had been made by EGS, on its own and Gulf Oil's behalf, and by APOC. But there was no suggestion of an offer for exploration rights on those occasions.

It was not until 1930 that matters took a positive turn. The position at that time is summarised by correspondence between the syndicate, the Political Resident, and the Colonial Office, then presided over by Lord Passfield. On 1 September 1930, the Political Resident in the Gulf wrote from Bushire to inform the Colonial Office that the Political Agent, Colonel Dickson, had informed Holmes that on no account would the Shaikh of Kuwait consider an application for an oil concession which did not contain a British "nationality" clause. On 16 September, Holmes himself wrote to Lord Passfield. He said that when he left Kuwait on 8 August he had informed the Political Agent that he intended to ask for a concession covering the state of Kuwait. The Political Agent, he said, had told him that there was no objection, subject to the incorporation of the Bahrain provisions. By 4 October, Lord Passfield was asking the India Office and the Petroleum Department of the Board of Trade for replies to his memoranda with which were enclosed copies of Holmes' letter. On 6 October, the Resident wrote to Lord Passfield who, as might be expected of the notoriously thorough Fabian, was taking a personal interest in the proceedings and conducting much of the correspondence in his own hand. Col Biscoe had the honour to inform the Colonial Secretary that the syndicate evidently intended to transfer a concession, if granted, either to the Bahrain Petroleum Company or to another subsidiary of Standard Oil of California. With that letter he attached a draft of the concession agreement which Holmes had submitted to the Political Agent, together with his own suggested amendments. The syndicate followed up with a protest to the Colonial Office about the delay that was being caused by insistence on the nationality clause. The syndicate claimed that it had been given permission to proceed in 1925 and had accordingly entered into an agreement with the Eastern Gulf Company on 30 November 1927. The Colonial Office had been furnished with details of this agreement in the following December. Thus "negotiations with the Shaikh were started, his signature was obtainable, the agreement virtually signed when the Political Agent intervened and announced that the nationality clause would be insisted upon". It was then recalled that in connection with the Bahrain concession prolonged discussions had taken place with the British government who had asked for the substitution of modified nationality clauses, set out in letters from the Colonial Office dated 16 September 1929 and 31 January 1930. "As these conditions were mutually satisfactory, we and our American friends had complete confidence that such conditions would also apply to the Kuwait concession."

The syndicate hoped that it could now proceed. It was, in the event, nearly a year before anything significant happened.

Holmes was not idle meanwhile. He had obtained permission from Shaikh Ahmad to drill a hole just behind the Dasman Palace. The Shaikh gave his authority to erect a small percussion drill and rig just outside the city wall and close to the seashore. After he had drilled for about a month an explosion occurred and the rig was damaged. To some observers that explosion might have given substance to Holmes' hunch that oil or gas, or both, were present. But three years were to pass before anyone gave credence to the New Zealander's belief and by then most had forgotten that it was he who set the whole thing in motion.

Towards the end of 1931, the Anglo-Persian Oil Company sent one of its leading geologists, Mr Peter Cox, to the area.

At much the same time, Colonel Dickson sent a confidential report to the India Office. It was dated 19 December 1931. In it he outlined the views of Major Holmes on the possibility of oil being found in Kuwait, Bahrain and the Gulf generally. He also included three sketch maps of the areas concerned. He remarked: "I forward this document with some diffidence since its conclusions may be common knowledge in the oil world." He went on to state the views which Holmes had formed from his extensive travels and investigations.

"In the first place Major Holmes gave it as his considered opinion that there was a close connection between the 'oil' and the numerous fresh water springs (all of which were warm, some very hot, to be found elsewhere in the sea off the North East coast of Arabia, on Bahrain Island, the mainland of Hassa, in the vicinity of Qatif and up the whole length of the Wadi al Miyah, which series also obviously stretched up through Kuwait where they weakened) and the country south of the Euphrates right up to the vicinity of Najd. This water, according to Major Holmes, coming from the Persian mountains and not from the highlands of Hejaz and Najd, passing under the Persian Gulf and the lower end of Iraq, at great depth, was eventually forced up again . . . by a sort of fault in the earth's surface which took the form, to use an easily understood simile, of a great rocky wall or cliff. . . . The same cliff formation was trying to help the oil to the surface, as witness the bitumen. . . . In spite of the fact that oil experts always pretend that surface oil indications never really meant anything, Major Holmes said that he had yet to find an oil geologist who did not carefully follow such indications or who did not find his eventual 'sprouter' in the vicinity of such indications. At Kuwait there is, according to Holmes, a surface ooze of oil at Bahra and also in the sea close to Ras al Abid near the southern border, as well as extensive bitumen deposits at Jebel Burgan."

In the course of the long, detailed and involved discussions that followed between the United States government, the Foreign Office, the Colonial Office, the India Office, the Political Resident in the Gulf, the Indian Government and the British Cabinet, no subsequent reference was ever made to this document. On 6 December 1931, the US Charge d'Affaires in London, Mr Ray Atherton, presented to the Foreign Office his case for an "open door" policy on the part of Britain, with particular reference to Kuwait, where the Eastern and General Syndicate, with the approval of the US government, wished to conclude an agreement with the Shaikh on behalf of American interests. The U.S. Ambassador at this time was Andrew Mellon.

The Foreign Office's proposed reply to Mr Atherton was sent to the Petroleum Department of the Board of Trade for routine approval. On 7 January 1932, the Petroleum Department suggested the strengthening of a paragraph regarding the application of another company (Anglo-Persian) for exploration rights. It also suggested that the Assistant Under Secretary of State at the Foreign Office, Sir Lancelot Oliphant, might point out to Mr Atherton that the syndicate sometimes adopted doubtful means to secure its ends. "You will remember that in connection with Bahrain, after the Eastern and General Syndicate had been given quite a good concession, they came along again before they had done any serious work and pressed to be given rights over the whole of the territory, thus seeking to exclude any other company . . . it looks as if they intend to do the same in Kuwait."

The Colonial Office also had strong views on the content of the reply which the Foreign Office should make to Atherton. It agreed to the draft reply, subject to the inclusion of the Petroleum Department's amendments. It added: "Mr Atherton might be informed that HM Government are not too favourably disposed towards the Eastern and General Syndicate." The Colonial Office also observed that the India Office was most reluctant to "abandon the attitude of maintaining the British control clause".

At this time the Foreign Office pointed out that an inter-departmental committee of the government had agreed that it was desirable to accommodate the Americans and that the attitude of the India Office was contrary to the conclusions of that committee.

The Colonial Office kept up its vigorous disapproval of the syndicate. "We are further inclined to argue . . . that in any further interviews between Sir Lancelot Oliphant and Mr Atherton, Sir Lancelot might suggest that the Eastern and General Syndicate are not really the sort of company that we like." By 15 January, draft A of the proposed reply to Mr Atherton had been finalised.

Meanwhile, the India Office maintained a rigid devotion to the control clause. It recommended the inclusion in a statement to the Americans of a report from Lt Col Biscoe, the Political Resident at Bushire, who had been sent to talk to Shaikh Ahmad. This report said that the Shaikh had told him (Biscoe) that he was "averse from dispensing with the British nationality clause and that the Shaikh does not regard himself as in any way committed by his letter to Major Holmes".

On 20 January, Sir Lancelot Oliphant wrote to Sir Robert Vansittart, the Permanent Under Secretary at the Foreign Office. "Both the US Ambassador and his counsellor have of late raised with the Secretary of State and me respectively the question of an oil concession in Kuwait in which an American company is interested. . . . One point is proving difficult of solution. The Americans have expressed the hope that the British nationality clause which was relaxed in the case of Bahrain, will not be insisted on in Kuwait. I assume you will agree that at present it is essential so far as possible not to be obstructive to American interests. The Colonial Office, the Petroleum Department of the Board of Trade, and the Foreign Office would all be willing to abandon the British nationality clause, but the India Office is extremely sticky. . . . May I tell them, on grounds of policy, it is desirable that we meet them on this point?" In a marginal note, Sir Robert observed: "Agreed, but don't over emphasise *policy*."

By now, draft A of the Foreign Office reply to Mr Atherton, had become draft B, with some substantial alterations. It showed, as was reflected in Sir Lancelot's letter to Sir Robert Vansittart, that the Colonial Office and the Petroleum Department had been won over by the Foreign Office. But the India Office, whose Minister at the time was Sir Samuel Hoare, remained adamant. Paragraph 2 of the heavily thumb-marked draft reply to Atherton bore the following India Office comment:

> "You will appreciate that the Shaikh of Kuwait is an independent ruler, although his special treaty arrangement with HM Government, whose protection he enjoys, led him to seek their advice on an important question of policy and (have) caused him many years ago to give them an assurance that he will (would) not grant an oil concession . . . without their consent."

On the same day that he wrote to his chief – Sir Robert Vansittart – Sir Lancelot Oliphant also communicated with the India Office. "I understand that the IO see difficulty about abandoning insistence on the British nationality clause and would like us in our reply to the American Chargé d'Affaires to state specifically that the Shaikh of Kuwait . . . desires to retain it. . . . Were we to tell the Americans that the Shaikh insists on the British nationality clause, they would scarcely believe that the Shaikh . . . is acting on his own initiative, but would assume that he is acting under pressure from HMG. . . . There will be some risk if, later on, the APOC drop out and we then maintain our insistence on the nationality clause and advise the Shaikh against granting a concession to Eastern and General Syndicate."

A week later, on 29 January, Mr Atherton presented himself at the Foreign Office. He was received by Sir Lancelot who was unable to give him a "categorical answer" regarding the Shaikh's letter, since the matter was being discussed with the India Office. The "Shaikh's letter" referred to was the communication of 2 July 1931 in which the ruler told Holmes that he must insist on the incorporation of a British nationality clause. The surviving documents of the period do not tell of Atherton's state of temper when he left the Foreign Office. Faced with continuing India Office intransigence, the Foreign Office devised a new approach – Sir Lancelot expressed anxiety about the safety of American citizens in the Kuwait hinterland. Atherton's reply, as minuted by the FO, did nothing to conceal the growing impatience of the Americans. There were, he believed, American drillers in the employ of APOC in Persia; where was the difference apropos Kuwait? He added: "I have reason to believe that the activities of APOC in Kuwait might be subsequent to my own representations of 6 December."

Sir Lancelot could not be sure, but would look into the matter.

The Foreign Office, like Mr Atherton, was becoming restive. The argument within British government departments was assuming the proportions of a first-class row. A meeting of the various ministries was called at which Sir Lancelot explained the difficulty of "sheltering behind the Shaikh", and the undesirability of causing the Americans unnecessary grievance. He said that he intended asking the India Office to telegraph the Government of India suggesting that the control clause be waived. The telegram was duly dispatched. Then, on 4 February, the Anglo-Persian Oil Company asked for government help in obtaining permission from Saudi Arabia to carry out a geological reconnaissance in Hasa and the Neutral Zone, so that it could reach a "definite conclusion" regarding oil prospects in Kuwait. On 10 February, the Consul at Jedda accordingly received instructions from the Foreign Secretary, Sir John Simon, to apply for permission on behalf of APOC. A touching footnote was supplied by the oil company. It offered to pay for any telegrams that might be necessary. Ibn Saud replied promptly to the British Consul. He would not grant permission to enter his territory. As for the Neutral Zone, he suggested that applications should be made to the Shaikh of Kuwait through the Political Agent.

On 23 February, the Foreign Office recorded another communication from Mr Atherton. It would, he said, be hard on American interests if British interests were allowed to obtain an advantage pending a decision about oil exploration in Kuwait. He, Atherton, had just heard that a large British contingent had arrived in Kuwait with machinery and other equipment. He desired to invite attention to the views which he had already expressed. On 26 February, these views were elaborated in a long memorandum which the Chargé d'Affaires left with Sir Lancelot. It underlined the US government's concern about the activities of APOC. "Not all their people in Kuwait had cleared out", it remarked.

It was now nearly three months since Mr Atherton had delivered his memorandum asking Britain to abandon the nationality clause.

Still, Sir Lancelot Oliphant's reply was passing from one Whitehall department to another, accumulating pencilled amendments on the way. It was becoming increasingly difficult to know exactly where the different parties stood. Towards the end of February, for example, the India Office was telling the Viceroy that "the balance of advantage lies in the admissibility of American interests" and that failure to exclude foreigners in Bahrain had "queered Kuwait's pitch", while the Political Resident suggested to the Secretary of State at the India Office that "the Shaikh of Kuwait be advised that his letter to Major Holmes is ambiguous".

Then, at the instigation of the Colonial Office, the armed forces were called in to the discussions. The Admiralty expressed itself with traditional singlemindedness. It let it be known that it would strongly resist any attempt to drop the nationality clause. Sir Lancelot Oliphant's draft reply to Atherton was scrapped and a new one attempted, this time stating: "The concern of HM Government is not so much with nationality of any interests involved . . . as with their duty to secure the best terms possible for the Shaikh of Kuwait." The Admiralty still objected.

On 11 March 1932, a meeting at the Foreign Office was attended by senior representatives of the armed services, the India, Colonial and Foreign Offices and the Petroleum Department. Sir Lancelot told the assembled company that the Indian Government had now overruled the Political Resident. In a message to the Secretary of State for India, the Viceroy's government had recalled that "as long ago as 1903 it had been laid down that the policy of excluding legitimate foreign trading interests from the Gulf was undesirable and took the view that it was inexpedient to maintain a nationality ban". The Admiralty was unimpressed.

On 14 March, still unable to finalise his formal reply to the Americans, Sir Lancelot decided to send a friendly letter to the Chargé d'Affaires: it began "My dear Ray," and went on "As the Eastern and General Syndicate is aware, APOC is also considering applying for a concession and indeed made a formal application to the Shaikh before the syndicate appeared on the scene, but the terms were unsatisfactory. The company have not lost interest, however . . . and are to carry out a geological survey . . . Sir John wishes me to say that he will be considering the whole matter in the near future and yet some further delay must be inevitable. We are so sorry."

On 15 March a memorandum from the Admiralty, classified secret, was delivered to the Foreign Office. It read: "I am commanded by My Lords Commissioners to forward for the information of the Secretary of State, a copy of a memorandum on the question of the application of the British control clause to this territory.

"It is, perhaps, unnecessary to emphasise the vital importance of bringing under British control at least sufficient potential output to ensure the mobility of our armed forces in war."

On the American issue, the Admiralty memorandum expressed general agreement with the Political Resident's views. It concluded: "In connection with this question the following quotation from Capt. Mahan of the US navy may be useful in discussion with the US Chargé d'Affaires.

"A concession in the Persian Gulf by arrangement, or neglect of local commercial interests, which now underlie political and military control, will imperil Great Britain's naval situation in the further East, her political position in India, her commercial interests in both, and the Imperial tie between herself and Australasia."

It was a well-aimed shot from Admiralty Arch, but it was unlikely to sway the ultimate issue. By 19 March a further Admiralty memo was prepared, reflecting the pressures that had been brought to bear in the intervening four days. It suggested the adoption of the policy laid down by the Petroleum Department in place of the British control clause. This would make it possible for foreign capital to be employed in the development of any concession "within the Empire", provided the parties concerned had no more than a 50 per cent holding in the operating companies. The Admiralty made further recommendations of its own. *Over* 50 per cent of the capital should be British. The company should be registered in Britain. At least 50 per cent of the crude oil should be refined on British territory and the plant should be capable of producing fuel oil suitable for naval use.

Sir Lancelot Oliphant began to redraft his reply to Atherton and, at the same time, to prepare a memorandum for the Foreign Secretary. He outlined the substance of the Admiralty's disapproval, as well as that of the Royal Air Force. Perhaps Sir John Simon could raise these issues in Cabinet?

On 30 March, Ray Atherton again took the initiative. In a note to the Foreign Secretary he said: "I have the honour to inform you that my government recalls the enquiry which it made through this Embassy in 1929 as to the policy of HM Government in the matter of the holding and operating of petroleum concessions by American nationals in British protected Arab territories such as Bahrain."

He went on to recall that the arrangement then agreed upon was "only just and fair in view of the extremely liberal treatment accorded in the USA and its possessions in reference to operation of petroleum concessions by British companies.

"My government had therefore supposed that the policy of HM Government would be no less liberal . . . in the matter of . . . Kuwait."

The US government understood that, contrary to the Colonial Office's claim, the Shaikh was quite agreeable to the specific entry of Eastern Gulf Oil Company, and to granting a concession without inclusion of the nationality clause. He went on to complain: "Despite promises at the time of the Bahrain concession, the Colonial Office later qualified its consent to American participation with the specific purpose of preventing entry into that territory (Kuwait) of the Eastern Gulf Oil Company. Continued insistence of Colonial Office . . . has seriously handicapped E & GS in bringing to conclusion negotiations with the Shaikh."

The situation was further complicated, said the note, by the fact that "at the very moment His Majesty's Government had under consideration the petition of the syndicate . . . permission was granted to the Anglo-Persian Oil Company, a rival concern, to send a small party of geologists to Kuwait." The US Embassy had repeatedly requested the Foreign Office not to permit Anglo-Persian to proceed with its operations "pending a decision . . . regarding open door rights for American nationals".

The note concluded: "The government of the US greatly regrets that no effect has been given to its requests in this matter, but would appreciate being assured by HM Government that this fact will not be allowed to militate against the position of the syndicate and its affiliate, Eastern Gulf Oil Company."

The Chargé d'Affaires was informed that the Foreign Secretary would be away in Geneva for a week but that he would bring the matter before the Cabinet immediately on his return. The American note, with an explanatory text from the Foreign Office, was circulated to the Cabinet on 4 April.

At this stage, prompted by ministerial enquiry, some germane documents were produced. The Colonial Office sent the Foreign Office copies of the correspondence between the Eastern and General Syndicate and the British government, seven years before. The relevant part of the letter from the syndicate, dated 1 September 1925, and signed by H. T. Adams, read: "My directors have no observations to make on the draft as it now stands, and would be pleased if you would forward a copy to the Political Resident in the Persian Gulf with the information that His Majesty's Government will raise no objection to our representative concluding an agreement with the Shaikh of Bahrain. . . . My directors take it that no objection would be raised by the Colonial Office to our representative concluding a similar agreement with the Shaikh of Kuwait (such an agreement to be based on the terms of the Bahrain Agreement)."

The reply from the Colonial Office, dated 3 September 1925 and signed by R. V. Vernon, read: "I am directed by Mr Secretary Amery . . . to inform you that the Political Resident in the Persian Gulf is being informed that HM Government would not raise any objection to the conclusion of your agreement with the Shaikh of Bahrain. . . . With regard to the question raised in the last paragraph of your letter under reply, I am to say that Mr Amery has no objection *mutatis mutandis* to the conclusion by your syndicate of an agreement in similar terms with the Shaikh of Kuwait, and the Political Resident in the Gulf is hereby instructed accordingly."

The Cabinet met on 6 April 1932. It was agreed (1) that the Foreign Office should ascertain through Sir John Cadman the position of the Anglo-Persian Oil Company in the concession question, and (2) that unless, as a result of the above consultation, new considerations were raised causing alteration of views or a need to consult the Cabinet again, the Foreign Office should be authorised to inform Mr Atherton that the inclusion of a clause insisting that the concession be confined to British interests would not be insisted upon, but that any application would be examined to ascertain what would best serve the interests of the Shaikh.

On 8 April, Sir Lancelot Oliphant prepared a letter to Atherton for Sir John Simon's signature. In a covering note he explained that although the Admiralty still insisted on the inclusion of specific safeguards, the Petroleum Department nevertheless believed that the most important safeguards were implicitly acceptable to the Americans and that "premature insistence on these provisions would make the Americans suspicious".

On the same day, 8 April, Sir John Simon wrote to the Prime Minister and the Chancellor of the Exchequer, Mr Neville Chamberlain.

To Ramsay MacDonald he explained that as he would miss the next Cabinet meeting, he would like to confirm the Prime Minister's agreement to a slight variation to the procedure decided upon by the

Cabinet. "The decision was that I should see Sir John Cadman first and write to Mr Atherton afterwards." Explaining that nobody had as yet been granted a concession and that, meanwhile, the Americans wanted an assurance on the nationality clause, Sir John announced that he intended to send a note to the US Chargé d'Affaires and "will arrange for Cadman to be informed immediately afterwards. The reason why this order of events is best is that Atherton has more than once hinted that we have been holding up a reply to him in the special interests of Anglo-Persian". He protested to his chief that such was not the case, but that it would be very undesirable to give the American Embassy any further grounds for "thinking we were being sly".

He offered much the same explanation to Chamberlain for the procedure to be adopted.

At long last, on 9 April 1932, the British government's reply to Mr Atherton was sent to the US Embassy. It summarised the many exchanges that had already taken place and took up Mr Atherton's claim that the Shaikh was agreeable to the entry of the Eastern Gulf Oil Company. "His Majesty's Government have caused enquiries to be made of the Shaikh, who replied that he was still averse from receiving in his principality a company other than an entirely British one. . . . It will be observed from a reference to the Shaikh's letter (to Holmes) that its final sentence only expresses a readiness to discuss the matter further with Major Holmes, after agreement has been reached between the syndicate and HM Government." Paragraph 4 of Sir John Simon's note concluded: "On a balance of all the conflicting considerations His Majesty's Government are, however, now prepared, for their part, not to insist in this case that any concession must contain a clause confining it to British interests, if the Shaikh for his part is willing to grant a concession without such a clause." The letter continued to stress the British government's belief that Anglo-Persian had a prior claim to negotiate, and that the syndicate's application in its existing form was unsatisfactory. It ended with a seventh paragraph stating the new stance of the government.

"The position therefore is that His Majesty's Government, for their part, are prepared to agree to the omission from any oil concession, which the Shaikh may be prepared to grant, of a clause confining it to British interests. If, therefore, the Eastern and General Syndicate desire to renew their application to the Shaikh for a concession, which they would subsequently transfer to the Eastern Gulf Oil Company, His Majesty's Government will raise no objection to the application being taken into consideration together with any other applications for oil concessions which may be forthcoming from other quarters."

After the die was cast, the Admiralty continued to insist on measures of British control. On the day before the Foreign Office reply was dispatched, the Secretary of the Admiralty was trying to substitute a formula which would not tie the Secretary of State's hand too precisely. But the Foreign Office and other departments concerned were pleased at least to have reached the stage of reply to the persistent Americans. The Admiralty's had become a lone voice of caution.

It remained for the government to report its decision to Sir John Cadman of the Anglo-Persian Oil Company and to the Political Resident at Bushire.

Sir John called upon Sir Lancelot Oliphant. In a note to the Secretary of State after the meeting, Sir Lancelot observed: "Today, in accordance with your instructions, I asked Sir John Cadman to call on me, when I explained that, but for your departure for Geneva, you would have had the pleasure of receiving him yourself." He then explained to the Secretary of State that he had told the Anglo-Persian chairman that HM Government no longer insisted on the British nationality clause, that future applications would be considered on their merits, and that the US government had been informed accordingly.

Sir John Cadman told the Under Secretary that, though his company had recently been showing considerable activity in Kuwait, this had been . . . in no way inspired by the fact of the Americans wishing to obtain a concession. In his own opinion, while there may be some oil in Kuwait, this would be *of a very heavy nature and not of a type which would, in the long run, be of interest to the Anglo-Persian Oil Company*. Sir Lancelot wrote: "He (Cadman) therefore told me for your personal information that the Americans are welcome to what they can find there." And he concluded: "I think from the political point of view, which is what primarily concerns the Foreign Office, that this is eminently satisfactory. But I see no need to inform the US Embassy at present of this withdrawal by the Anglo-Persian Oil Company – time enough later, when we have heard about the terms of the American draft concession."

Sir John Simon appended a postscript to the note: "I agree entirely. Let the PM know that all works out according to plan. Chancellor of Exchequer might also like to know."

On 26 April the Colonial Office began to draft a telegram to the Political Resident, Colonel Biscoe, as well as a letter to the syndicate. At the same time it put on record a further discussion with the head of Anglo-Persian. "Sir John Cadman called here last Saturday. . . . He said that he had given instructions for the Anglo-Persian Company's geologists in Kuwait to be withdrawn for the moment. When asked whether this meant that the company had lost interest in Kuwait, he said that this was not so. They were waiting for the result of certain tests . . . and gave us to understand that the company would be making proposals regarding Kuwait in the near future, either to the Colonial Office or locally."

The telegram to Biscoe outlined the details of the letter to the Americans, a copy of which had been airmailed to him on 22 April, and told him that he should be guided by the provisions of that letter in any future negotiations. He was asked to explain to the ruler that as regards the application of the Eastern and General Syndicate, "His Majesty's Government for their part, do not insist upon the insertion of a British nationality clause and that he is now at perfect liberty to dispense with the clause or not as he feels best. . . . HM Government do not consider that they could properly advise the Shaikh to give prior and preferential treatment to the syndicate, but hold it necessary that any application for a concession from any quarter be examined with a view to deciding which, if any, will best serve the interests of the Shaikh."

The letter to the syndicate caused another interdepartmental wrangle. After recapitulating the British position with regard to the Shaikh's interpretation of the 1899 agreement and the syndicate's interpretation of its talks with the Shaikh, it went on to repeat the phrase about dropping British insistence on the nationality clause, "if the Shaikh, for his part, is willing to grant a concession without such a clause . . .". The India Office suggested that the sentence should begin with the word "Whether" and continue with "is a matter entirely for His Excellency's decision." It was the word "entirely" in this amendment which brought further wrath from the Admiralty. The new approach, they said, showed a total concern for the Shaikh's interests and none for Britain's strategic position. Such arguments served only to delay still longer the instructions which HM Government should properly have given to its local representatives at the earliest opportunity. The telegram, as amended, eventually arrived at Bushire on 9 May. While procrastination marked the proceedings in London, Major Holmes was busy in Kuwait.

On 10 May 1932, Lt Col Biscoe wrote to the Colonial Office, enclosing a letter he had received from the Political Agent in Kuwait. In view of all that had gone before, it was a truly remarkable correspondence that now emerged, and none of the letters involved was more remarkable than the Political Agent's, written on 1 May.

"I have the honour to inform you", said Dickson, "that Major Frank Holmes arrived from Bahrain on 27 April and dined with His Excellency the same evening. Major Holmes called on me the following day and I returned his call on 29 April."

The Political Agent went on to say that Holmes on arrival had made an unnecessary set speech in praise of Mr Mellon, the American Ambassador, who according to Holmes was the most popular man in England at that time, and that he had made a gratuitous attack on the Political Resident and British government officials, whom he styled "out of date persons who think of nothing else but how to maintain their personal power over local potentates". Col Dickson ingeniously reflected Holmes' manner of speech and added the rider "He graciously excluded me from his remarks."

Dickson then described how, with an air of childish triumph, Holmes called his clerk who produced a telegram received from the syndicate's London Office. It quoted the British government's note to the US Embassy and underlined the phrase "if therefore the Eastern and General Syndicate desire to renew their application to His Excellency the Shaikh of Kuwait for a concession, *to be subsequently transferred to the Eastern Gulf Company,* His Majesty's Government will raise no objection . . .". Holmes then handed Dickson the telegram in duplicate, the original of which the Political Agent forwarded to the Resident in Bushire. In doing so, he made his views on Holmes' latest salvo very evident.

"Major Holmes makes no mention of the fact that the syndicate received their so-called extract of a Foreign Office letter from America, but on the contrary states that His Majesty's Government have been in direct communication with the syndicate.

"The clever part of Major Holmes' letter to me, which . . . he must already have shown to the Shaikh, lies in the fact that the Shaikh must now think that Holmes and his backers are more powerful than he ever gave them credit for, and that they have forced HM Government to go back on their original decision to insist on the nationality clause. This I submit will not only shake the Shaikh's faith in us badly, but knowing him as I do will certainly set him wondering whether he can really count on that assistance . . . which he hopes and expects to get in other important matters." The Political Agent goes on to quote some characteristic but hardly worrying Holmes' excesses. "A wonderful victory", "the most abject climb down that it had been his fortune to see". What did, apparently, astonish him was "the way the Admiralty had raised no objection". Dickson was obviously in two minds about Holmes' reaction to the words, omitted in an earlier discussion, "to be subsequently transferred to the Eastern Gulf Company". Holmes it seemed did not approve of the words, maintaining that though the money for developing the concession might come from America, it had always been his intention that a British company should operate it. " I found this very strange, especially when he turned round and begged me not to divulge (to the Shaikh) the underlined portion of his telegram, or what he had said about it to me." Dickson thought that this nervousness on Holmes' part might indicate that there was a break between him and his principals, or that he was simply posing as the champion of the Shaikh's interests (assuring him, in effect, of an American-backed concession while retaining British protection), or that, as often happened, he was just acting.

After some apparent misgivings at the "smart" and "shady" way Holmes got hold of the earlier letter from Shaikh Ahmad expressing a willingness to negotiate an agreement with the syndicate, Dickson made a point that was to cause a great deal of head shaking in the Gulf and in London.

"If it is an actual fact that His Majesty's Government have also told Major Holmes' company, the Eastern and General Syndicate, London, that they can go ahead now without a British nationality clause, the only course would appear to be to tell the truth to the Shaikh and to allow Major Holmes to reopen negotiations for his oil agreement. Both parties however should be informed officially, and for preference direct by you. I think that HM Government still reserve the right to examine and approve the actual terms of any agreement . . . before it is signed by either party. In this way only can we ensure that the Shaikh gets a fair deal, does not sign away his birthright, and that our interests are safeguarded."

On 29 April 1932, Holmes wrote to the Political Agent in Kuwait:

"I have now been instructed by my London Office to renew on behalf of the Eastern and General Syndicate Limited the application to His Excellency Shaikh Sir Ahmad al Jabir al Sabah KCIE, CSI Ruler of Kuwait, for an oil concession within His Excellency's territory. Further, I have been informed that His Britannic Majesty's Government have written to my principals as follows . . .". He then quotes the British government's words to the American Embassy regarding the omission of the nationality clause. "I take it that HM Government have informed the Honourable the Political Resident in the Persian Gulf of its decision. . . . I wish to renew my company's application without delay, and would be pleased if you would kindly inform His Excellency . . . officially by letter, that HM Government has agreed to the omission of the caluse . . .".

By early June, the indefatigable Holmes had sent a draft concession agreement, his third attempt to date to the British Government. It was forwarded to the Colonial Office by the Resident on 7 June.

IN THE NAME OF GOD THE MERCIFUL. THIS AGREEMENT made . . . at Kuwait between His Excellency Shaikh Sir Ahmad al Jabir al Sabah, KCIE, CSI, Ruler of Kuwait in Arabia (hereinafter called THE SHAIKH which expression where the content so admits shall include His Heirs, Successors, Assigns and Subjects) of the first part, and Frank Holmes of 18 St Swithin's Lane, EC4, London, England, the true and lawful attorney of the Eastern and General Syndicate Ltd. . . . WHEREAS THE SHAIKH is desirous of developing the Oil and Petroleum Resources of His Territory . . .".

There followed twenty-four clauses which may be summarised as giving the syndicate exploration rights over the whole of Kuwait except for the old walled city and sacred buildings, over a period of seventy calendar years, the exact delineation of the conceded territory to be decided within the first five years.

Exploration was to begin within nine months. There followed proposals, customary in all oil concessions, relating to the erection of plant and installation, employment of local labour, freedom of movement of personnel and products, the use of communication and transport facilities. Then to the meat. The Shaikh would be entitled to a customs duty of one per cent on all oil and associated products, calculated on the value of the oil "at the wells producing same". The company was to be exempt from all harbour duties, taxes, imports and charges, but the Shaikh would have the right to collect duties from ships other than those acting for the company. Then clause 8: "If the ownership of this concession be transferred to another company and THE SHAIKH undertakes to sanction such transfer when so requested, PROVIDED . . . (a) that the assignee company shall be one organised or registered either in Great Britain or organised or registered in Canada and shall maintain an office in Great Britain which shall at all times be in charge of a British subject, (b) that of the five directors of the assignee company one director shall at all times be a British subject . . . persona grata to His British Majesty's Government, (c) that the assignee company shall at all times maintain in the region of the Persian Gulf an official to be called the CHIEF LOCAL REPRESENTATIVE . . . approved by HM Government. Within sixty days of signing the Agreement, the Company would pay to the SHAIKH Rs30,000, and on each subsequent anniversary Rs20,000." Should the company find oil in commercial quantity, these annual payments would stop and the Shaikh would receive instead a royalty of three rupees and eight annas per English ton of net crude oil, excluding that required for local use. The amount payable to the Shaikh "shall not be less than Rs70,000 in any complete calendar year in which THE COMPANY continues to work". On the termination of the agreement, all installation, buildings, plant and machinery would become the property of the Shaikh, unless the concession was discontinued for two years continuously or through an inability to maintain due payments, within a period of thirty-five years, in which case the company could claim its plant and machinery. In the event of discrepancies between the English and Arabic versions, the English would prevail.

The Eastern and General Syndicate sent a further copy of the draft to the Colonial Office on 10 June, with a request for an early reply.

There followed another bout of interdepartmental exchanges which traversed a lot of old ground and, using Holmes' latest draft agreement as the starting point, some new territory.

In a letter of 10 June from the Political Resident, Holmes again came under attack. "I do not consider (him) a particularly desirable person, and shall be glad to see him replaced by someone else in Bahrain when occasion offers. In any case, he should not be the Local Representative in Kuwait." Later in the letter he remarked: "I regard the discussions out here as very unsatisfactory, as the Shaikh is a child in the hands of Major Holmes, while neither Colonel Dickson nor I possess any technical knowledge."

There can be no doubt that the British government departments were thrown into confusion by the prompt response of Holmes to their own acquiescence in American demands for an "open door". Wariness marked every government move from now on. And the effort to persuade Anglo-Persian to take unequivocal interest in the proceedings becomes evident from the marginalia of the Foreign Office at this time. "The Shaikh has succeeded in making APOC abandon their shilly-shallying" was a fairly typical comment. By July, the Petroleum Department was in earnest communication with the Political Resident, and Anglo-Persian was making positive moves. In May the company offered £2,000 to the Shaikh for an option to prospect for eighteen to twenty-four months, but this was refused. The ruler wanted a definite proposal for a concession.

"This is an awkward business," began a Foreign Office memo of 26 July. Suspected by the Americans, prodded by the Shaikh, and confronted with a half-hearted approach by its own oil industry, Britain was heavily sandwiched. The unfortunate permanent officials of the Foreign and Colonial Offices, and the Petroleum Department, were hard put to it to cope with an obdurate Admiralty. Frank Holmes and the Americans were beginning to exert back-breaking pressures. The gloom was somewhat relieved by August, however, when the Petroleum Department was able to announce that Anglo-Persian had submitted a definite offer to the Shaikh. The government could let the two draft agreements go forward without interference and hope that APOC would win. The Colonial Secretary, now Sir Philip Cunliffe-Lister, was receiving via the India Office an almost daily bombardment of news and views from the Political Resident at Bushire. There was a copy of a letter, marked "personal" and dated 29 April, written to the Political Agent on the notepaper of the Athenaeum by A. T. Wilson (Sir Arnold Wilson). It suggested that Dickson would be disappointed that APOC had withdrawn its geologists – "for tactical reasons" – but that he should take heart from drilling that was going on around about. "We are at this moment engaged in putting down a deep test in the hills south of Bushire, and our observations at Kuwait have to be correlated with what we are getting in that well." He went on to suggest that APOC wanted to buy time – "though I do not think the company would be disposed to go higher than £2,000" – in order to complete deep well tests on the other side of the Gulf and correlate the information with that obtained from Kuwait, Iraq and Persia.

Letters filtering back from Bushire also showed that Chisholm had been to the Residency in May and had called on Dickson in Kuwait. He had not sought a meeting with the Shaikh. The burden of Chisholm's remarks to Biscoe seems to have been that his company would like an exploratory licence but did not wish the pay the Shaikh "more than about £1,000 a year". Hugh Biscoe, knighted in that year's honours list, made it clear that the Shaikh was not unnaturally very dubious about their intentions. In June, following Chisholm's visit, Mr N. A. Gass, deputy General Manager of APOC (later Sir Neville Gass, chairman of BP), went to Kuwait. The latter wrote to Sir Hugh from Abadan on 25 June, explaining that he had two long interviews with the Shaikh who had gained a "very bad impression" from the company's sudden withdrawal of its geologists in the previous April. Nevertheless, the Shaikh had sent a friendly message to the company and to Sir John Cadman. When Gass reported to London on his conversations with Shaikh Ahmad, Sir John Cadman gave immediate instructions for a concession agreement to be prepared. "We shall, of course, do everything possible not to keep HE waiting." A fresh breeze was discernible.

Mr Gass confirmed the company's wish to enter into formal negotiations in a letter to the ruler dated 25 June 1932. While the Anglo-Persian Oil Company prepared its first draft, the interested departments of the British government started to examine the syndicate's third effort. Another complication arose at this stage with the death of Sir Hugh Biscoe. He had been in close contact with the negotiations and it was not a particularly good time for someone else to have to familiarise himself with the facts and procedures. By 16 August, however, a new Resident, Lt Col T. C. Fowle, was able to advise the Colonial Secretary of a draft concession submitted by Anglo-Persian to His Excellency the Shaikh of Kuwait. Colonel Fowle had been informed that Mr Chisholm would arrive in Kuwait on 14 August to present the draft to the ruler. This gave financial detail in sterling and was subsequently revised to incorporate monetary detail in Indian currency. In its revised form, this initial draft provided for exclusive rights to explore the whole of the territory of Kuwait (excepting only the town of Kuwait and graveyards) to search for petroleum and cognate products. An immediate payment on signature of Rs50,000, plus a further Rs20,000 at an unspecified date in 1933, was offered. For a further period of two years from a date to be agreed in 1934 the company would be empowered to drill in any part of the territory to a maximum depth of 4,000 feet, and at the inception of this drilling programme a further Rs25,000 would be payable by the company. At the expiry of this two-year period in 1936, the Shaikh would undertake to grant a seventy-year concession to search for and produce petroleum, and to refine within the territory of Kuwait. At that time, and on each subsequent anniversary, the company will pay to the Shaikh Rs25,000, or royalty as set forth, whichever was the greater. The royalty would amount to two rupees and ten annas per ton. As soon as petroleum was confirmed in commercial quantity and quality, the annual payment to the Shaikh would be Rs50,000 plus royalties, with a minimum payment in the first year of Rs65,000, in the second year Rs80,000 and thereafter Rs100,000 yearly. In the case of unavoidable interruption of production, payment would reduce to Rs25,000 annually. There followed the usual provisions such as were incorporated in the syndicate's draft. With the approval of the Shaikh, and acting on the advice of the Political Resident, the company retained the right to transfer the agreement to any British company.

Mr Chisholm and his colleagues had studied the syndicate's terms well. There was little to choose between the bids as far as the ruler was concerned.

At this stage, the US Embassy became active once more, requesting to know why the British government was delaying its response to the syndicate's application and demanding that the Anglo-Persian Oil Company be given no special treatment. The Ambassador himself, Mr Mellon, was now handling the matter.

In order to be quite clear as to the position of Anglo-Persian, the Colonial Office attached to its copy of APOC's draft agreement, a letter written in 1925 and hitherto mentioned only in passing.

On 24 March 1925, the Colonial Office had written to the Anglo-Persian Oil Company: "Gentlemen, with reference to your letter of 18th February in regard to your negotiations for the grant of oil concessions in Bahrain and Kuwait, I am directed to inform you that . . . the Secretary of State regrets that he does not see his way to grant any extension beyond 31 March next of the period during which HM Government will recognise your priority of application . . .".

On 14 October 1932, a Foreign Office note recorded that Sir John Cadman, who had already had a talk with the president of Gulf Oil in London, was shortly leaving for America where he intended to approach the Mellon Group. Sir John felt that the group's vicarious activity in Kuwait "may be held to be an infringement in spirit of a sort of self-denying ordinance between the big oil companies not to overlap into each other's zones". Sir John was told that the government's concern was to ensure that Anglo-American relations were not disturbed in any way.

The day before, Mr Mellon had asked for an early appointment with the Secretary of State at the Foreign Office. The Ambassador, in fact, called on the 18th. He "manifested some impatience at the length of the delays," according to Sir Robert Vansittart. Mr Mellon was informed that the position remained as described in the note to Mr Ray Atherton, which explanation did not seem to pacify him greatly.

Sir Robert Vansittart wrote: "He (Mellon) says the matter has been on the stocks for over four years – I believe he is right in saying this – it is hard to think the Americans have no cause for complaint." On 2 November the Ambassador again complained that despite promises to the contrary, no acceptable communication had been made to him or to the syndicate.

The Foreign Office was clearly embarrassed and annoyed. "Nothing can make this case look well. Get the letter to the Petroleum Department off today. I shall almost certainly have Mr Mellon back next week, a little more acid", wrote Sir Lancelot. He added, "When he hears – as I suppose he will – that we have most impartially, and oddly enough, found that the APOC offer is slightly the more advantageous one, I think Mr Mellon would let me have 20 to 1 on that, and I certainly would not take the bet."

The American Department of the Foreign Office became understandably sensitive about the accusations to which it was being left open. "The Ambassador was perfectly polite throughout, but it was quite plain that he considered that we were acting with deliberate dilatoriness and indeed with duplicity," complained an angry Sir Robert, "and from what I know of this case, I should, in the Ambassador's place, feel exactly the same thing." He went on: "No one could possibly think otherwise than that these (delays) were being created for the benefit of the Anglo-Persian who refused their chance six years ago, and now wish to pop in a change of mind. And this I take to be approximately the truth. . . . Be this as it may I wish to make it quite plain to all departments concerned that I will no more submit to being placed in this intolerable position; nor will I accept any further excuse whatsoever for any further delay whatsoever . . .".

He concluded that particular memorandum: "Apart from this, for reasons of considerably higher policy, I do not wish to have acrimonious disputes with the USA at this moment. If, therefore, this matter is not immediately dealt with – and by immediately I mean by the end of this week – I shall raise very considerable trouble."

The British were in the position of brokers with a suspect reputation. There was only one outcome that could save the government's face, an agreement between APOC and the syndicate to share the concession and to negotiate jointly. Meanwhile, the other departments, particularly the Petroleum Department, began to look for plausible explanations for a delay which had become very protracted indeed. Again, they fell back on Holmes, that "slippery customer", and on an attempt to argue that the syndicate had not been given the green light back in 1925.

Stung into comprehensive reply by a letter from Vansittart which repeated much that was contained in his internal memorandum, the Petroleum Department went over all the old ground on, of all days, 5 November. It said that it was sorry about the embarrassment caused by delays but, stoking an already well fuelled fire, added that this was the more regrettable as a stern fight between Anglo-Persian and the syndicate was likely, and that a fairly firm stand might be required against the US government. The unhappy officials of the Foreign Office, helped by the Colonial Office and Petroleum Department, began the weary task of framing an official reply to Mr Mellon. On 11 November, Vansittart wrote to Mellon: "I am glad to be able to let you know that the department concerned have now completed the comparative examination of the draft concessions . . . and the document embodying the result of this examination is on its way to the British authorities in the Persian Gulf. . . . Meanwhile, I am arranging to have a detailed reply prepared to the various other points raised in the memorandum which you left with me on November 2nd."

"Dear Sir Robert, I am glad to have your letter . . . and I am awaiting with interest the detailed reply to the memorandum to which you refer," replied the Ambassador.

Sir Robert's reply went over the heavily worn ground of the respective applications of APOC and the syndicate, and contained repeated assurances that HM Government had no wish to cause any avoidable delay or to interfere in any way with the Shaikh's freedom of choice between the two applications. It was sent on 23 November. Meanwhile, the Petroleum Department was busy preparing a comparative appraisal of the two draft agreements. It put all the clauses under the microscope and decided that there was little to choose between them, except perhaps in financial terms where it found the APOC offer slightly more generous. The British approach had become one of leave it to the Shaikh. It was summarised in a Foreign Office comment of 12 November on a note from the India Office: "It is suggested that in view of strong pressure exerted by United States' interests, the burden of decision between Eastern and General Syndicate and Anglo-Persian Oil Company should be laid on the Shaikh of Kuwait." The Admiralty remained decidedly unhappy.

Throughout December 1932 and January 1933 a long and complicated dialogue took place between London and Bushire. Sir John Cadman returned from his talks with the Mellon Group in America and gave the Foreign Office to understand that both sides would continue to press for the concession. In a letter to Sir Lancelot Oliphant dated 31 December he made an evocative aside: "In the meantime, there is nothing concrete but, believe me, we are in full cry for the Kuwait concession and shall press to the bitter end, unless we are able to fix things up with the Mellon people." At about this time the government in London received from the Resident a copy of the new draft agreement containing "certain amendments and revisions", which the General Manager of APOC, Mr E. H. O. Elkington, had forwarded to the Political Agent. It suggested increased down payments to be made to the ruler, including a royalty from natural gas, and naming sacred buildings among the inviolable areas. It also granted the Shaikh the right to lease mineral rights other than oil.

On 14 January 1933, the Foreign Office was informed that the ruler had examined the comparison of drafts prepared by the British government and had been informed by the Political Resident that formal negotiations could begin with both parties. Holmes was in London at this time. On 19 January, the India Office expressed its interest in correspondence between Oliphant and Cadman, and asked for more information on a proposal that negotiations might continue on the basis of a fifty-fifty share between APOC and the Mellon Group. It was pointed out that "heavy weather" had been made of convincing the Shaikh of the need for safeguards. "Were they now being asked to foist an American company on the Shaikh?" Anglo-Persian, meantime, had prepared three alternative proposals to put before the ruler, while the Lord Commissioners of the Admiralty fired another broadside suggesting that if any deal were to be done between Anglo-Persian and the Mellon Group, the Americans should offer compensating concessions to Britain; in particular that the Bahrain fields should be transferred to the Burmah Oil Company, that any Anglo-American agreement should embrace Hasa as well as Kuwait, and that in any such agreement British influence must predominate.

By 14 February a clearer statement of Sir John Cadman's discussions with the Mellon Group had become available. APOC, it seemed, were determined to obtain the concession but hoped to make a bargain with the Americans giving it certain advantages in the USA in return for a 50 per cent shareholding in a subsidiary company to be formed in Kuwait. A complication in the Anglo-Persian Company's position became apparent in a letter from George Rendel, who was with the British delegation to the League of Nations in Geneva, following a meeting with Sir John Cadman. "The United States group appear to be afraid that if the Anglo-Persian Oil Company . . . secured the concession . . . they might refrain from developing it. . . . If they secure the Kuwait concession now, and develop it at once, the Persian government may revive the accusation that they are frittering away their energies elsewhere than in Persia." Colonel Dickson at about this time reported that the Shaikh was refusing to be hurried in his decision but appeared to be leaning towards the Eastern and General Syndicate because he, too, had doubts about the willingness of APOC to work the concession. Sir John Cadman had another irritation to cope with. In the event of the British and American companies going it together Holmes, or at any rate EGS, could claim a royalty of one shilling a ton according to the terms of the agreement between those parties of November 1927. "Little short of monstrous", Sir John was quoted as saying. Towards the end of February the American Ambassador in London announced his recall and Sir Robert Vansittart began to prepare for a farewell visit. The Permanent Under Secretary was assured that a deal between APOC and the Mellon Group was now a virtual certainty since both sides had said that if they were awarded the concession they would cut the other in on a fifty-fifty basis. According to the Political Agent, nevertheless, the tide had turned against APOC because of the Shaikh's grievances with Britain over the Saudi blockade of his country and the failure to exempt his date gardens in Iraq from taxation.

On 2 March, the Colonial Office reported to the Foreign Office that Sir John Cadman had asked whether the government could bring pressure to bear on the Shaikh to give the concession to Anglo-Persian since his company had come to an agreement with the Gulf Company that if they obtained the concession they would work it and Gulf would have a 50 per cent shareholding. The government maintained that it could not influence the ruler in any way. In the same month, Sir Samuel Hoare, Secretary of State for India, considered the possibility of inviting the ruler to London, thus counteracting a similar invitation already extended by Holmes on behalf of the syndicate. After discussion with the Foreign and Colonial Offices it was decided that such a move would be open to misconstruction by the Americans. At this point, the Anglo-Persian Oil Company requested urgent government intervention in the matters of the blockade and the date gardens in order to enhance its negotiating position. It now seemed, following a conversation between Colonel Dickson and Shaikh Abdulla al Salim, that Holmes was promising Shaikh Ahmad American help in raising the Saudi blockade and in restoring his rights over the date gardens. In a letter to Shaikh Ahmad of 12 February, the Political Agent suggested that His Excellency should avoid any detailed discussion with the APOC or syndicate representatives until he had decided to which company he intended to grant the concession. "If in doubt, Your Excellency should ask my advice . . . keep me, your friend, generally informed of the progress . . . of discussions with the representatives of the two oil companies."

In his reply the ruler said: "Please be so good as to inform the Hon'ble Political Resident that we shall be very careful to avoid the difficult question mentioned . . . we feel confident that His Majesty's Government has no other aim than the prosperity and welfare of ourselves and our country." He went on to tell Dickson: "Your Excellency may be quite assured that I rely on you, my friend, more than on myself and of course should I have any doubts . . . I would consider it most essential that Your Excellency be approached before they be definitely settled." That at least was a good sign for the Foreign Office.

In March, Sir John Cadman visited Kuwait and had a long meeting with Shaikh Ahmad at which he offered "quite generous terms", according to the Foreign Office, in return for a concession. The Shaikh replied that he must consult with Holmes before giving an answer. The Foreign Office could see nothing for it but to ask the Viceroy to allow the acting Political Resident, Colonel Fowle, to go to London for secret consultation. It did not want to publicise such a visit and thus invite Holmes to strike while the iron was hot. In April, Imperial Airways were asked to facilitate the speedy transport of Fowle to London. But, wisely, it was decided in the end that he should visit Kuwait first, thus relieving the trip of an element of suspicion.

Dickson, meanwhile, reported that the ruler had received an important communication from King Ibn Saud on 19 February and that he (the ruler) had subsequently made two secret journeys to the south. The first was on 20 February and the second, in a violent dust storm, on 27 February. Both journeys started late at night, said Dickson. On his return, the Shaikh announced that he had been on hunting trips. Dickson commented that he had taken with him Shaikh Abdulla al Salim, Shaikh Salman al Hamud and Shaikh Ali al Khalifah, all cousins, as well as his son Muhammad, Yusuf al Salim, and Holmes' confidential agent, Muhammad Yatim. The Shaikh never takes members of his family on hunting trips, observed the Political Agent.

Dickson's feeling was that the ruler had met Ibn Saud, at least on the second occasion, and that they had discussed the King's urgent need for money – and oil. As an after note, he added: "The Shaikh may have met the King's brother-in-law, Saud al Arafa, who is said to have returned from Mecca." He enclosed with his report, copies of fraternal exchanges between the King and the Shaikh. On 22 March, Colonel Dickson sent a further report on Holmes' movements to the Resident. Shaikh Ahmad, it appeared, had mentioned casually at a dinner party that Holmes, who was then in Bahrain, had prepared a new draft agreement. This turned out on further investigation to be two agreements, one covering an area of 1,200 square miles, the other 400 square miles. Separate payments would be made for each of the two concessions. The Shaikh told Dickson that in his view this was a vastly better offer than the Anglo-Persian one. Dickson protested that there could not be two concessions in such a small state. The Shaikh insisted that there could. "I think," said Dickson, "that the Shaikh got hold of the idea of a twin agreement from Ibn Saud." He suggested that the Shaikh would probably offer the larger area to the British.

Fowle duly arrived in England, and a large gathering at the Foreign Office heard his views on the concession situation, on the possibility of new discussions on the Neutral Zone coming up, and on such side issues as the syndicate's invitation to the ruler to spend the summer in England. Shaikh Ahmad had asked Dickson to obtain the British government's view of such a visit.

On 14 May Mr Chisholm informed the Political Agent that the

Shaikh wished to suspend the negotiations and would let both parties know when he was ready to resume them.

Frustration was now the dominant mood and a Foreign Office official perhaps put the general feeling into a nutshell when he commented: "The whole thing gives me an uncomfortable feeling that much is going on *sub rosa* of which the Political Agent is being studiously kept in ignorance." Perhaps so, but it was surprising what Dickson did know, or at any rate found out. He knew, for example, that Khan Bahadur Abdul Latif Bin Abdul Jalil, late director of customs, had returned quietly to Kuwait after leaving "under a cloud" of some three years earlier to seek exile at the court of Ibn Saud. "The worthy Khan Bahadur," said Dickson, "had returned with a free hand from the Saudi King to push the royal interest and that of APOC if it should benefit him (the King) financially." He appeared very anti-American and anti-Holmes.

The Khan Bahadur confirmed to Dickson that the Shaikh of Kuwait had journeyed into the desert to meet the King on 20 February to prepare the ground for the submission by Holmes of a draft concession for the Neutral Zone. He argued that Ibn Saud would wish Kuwait's concession to go to the syndicate, along with that for the Neutral Zone, since "obviously both must go together". Following a request from Holmes to be received by the King, Ibn Saud invited him to Jedda by telegram on 18 March. Abdul Latif stressed that either HM Government must bring pressure to bear on Shaikh Ahmad for a speedy settlement in favour of APOC or the British oil company must forestall Holmes in the Neutral Zone.

Dickson took the view that although Holmes, whose relations with the Arabian leaders tended to blow hot and cold, was *persona non grata* with the King, the latter nevertheless might use him to stress to the world the strength of Saudi influence in the area. Dickson suggested also that the 400 square miles in the "twin" proposal put forward by Holmes referred to the Neutral Zone. The Political Agent's belief at this stage was that Shaikh Ahmad was doubtful of APOC's intention to work the concession and that he favoured the syndicate on grounds of practicality.

As he was preparing to send his report of the meeting with Abdul Latif to the Political Resident, Dickson received a call from Chisholm. The latter had decided that he would ask the Shaikh that day, 27 April, for a letter to Ibn Saud to the effect that if he (Shaikh Ahmad) gave the concession to the Anglo-Persian Oil Company, he would also . . . raise no objection to APOC taking out a concession for the Neutral Zone, and trusted that Ibn Saud for his part would agree also.

At this stage, a special service officer, Flight Lt E. Howes, stationed at Basra, came into the picture. From a letter dated 17 April sent by Dickson to this officer, it becomes apparent that the British government had pieced together a detailed intelligence report on the movements of the principals in these discussions. Ibn Saud's desperate need for money seemed to be playing an increasingly important part in events. Dickson told Howes:

"On 18 March Ibn Saud wired to Holmes to come to Jedda at once if he wished to be a bidder for the Najd oil concession." Holmes, Lombardi (a Standard Oil director) and Muhammad Yatim (Holmes' assistant or "Jackal" as Dickson described him elsewhere) flew to Cairo on 19 March. Holmes and Yatim went on to Jedda about 24

March. "You heard, no doubt, that last month Ibn Saud got £10,000 out of the Shaikh of Qatar and has now asked the people of Hasa to pay a year's revenue in advance."

A confidential report dated 30 May from E. H. O. Elkington in Abadan to Mr A. C. Hearn at the Anglo-Persian head office in London was optimistic. He revealed that Holmes had pretended to know nothing about the Shaikh's wish to suspend negotiations. Confronted by Dickson with a letter from the Shaikh stating that he had suspended talks with both companies, Holmes apparently launched into a vituperative attack on the probity of HM Government and APOC. He also asserted the power of the Americans to defend him against "their unjustifiable methods". Eventually, Dickson asked the ruler for an explanation of Holmes' ignorance of the position. "There is no doubt that Holmes' attempt to bluff . . . and to involve the Shaikh . . . has offended the Shaikh very deeply, and has further made him very suspicious . . .". It was a short-lived storm. Nine days after the Shaikh's decision to suspend the talks, on 23 May 1933, both negotiators were told by their respective principals to discontinue their efforts to obtain a concession, and not to attempt to persuade the Shaikh to reopen negotiations. Other plans were in embryo.

Altogether APOC was pleased with the turn of events. Mr Elkington also said: "The position therefore would appear to be eminently satisfactory, with every reason to believe that it will develop in a manner which will obviate the necessity of any American participation." Sir John Cadman showed the Foreign Office the letter with evident satisfaction on 9 June.

On 28 May, the ruler of Kuwait had written to APOC's chief negotiator: "I hope you are well and enjoying sound health. I have pleasure in acknowledging receipt of your friendly writing of 26th instant. I am exceedingly delighted to go through the contents of the wire dispatched by our friend Sir John Cadman, for which I sincerely thank him. . . . We trust, God willing, that this friendship will be a strong and permanent one, everlasting in amity and sincerity. I wish you good health and success." This note referred to the successful outcome of APOC's re-negotiation of the Persian oil agreement, of which Sir John had informed Shaikh Ahmad. Thus, APOC was wearing a broad smile in its dealings with the ruler despite the fact that the latter had called off the talks while it, likewise, had told its negotiator to hold off.

On 19 June, the Foreign Office recorded the fact that Standard Oil of California had obtained the Hasa concession and that the Iraq Petroleum Company looked like getting control of Qatar. Sir Samuel Hoare at the India Office stressed the increasing importance of Britain securing the Kuwait concession.

Towards the end of June, the British government was informed that the Shaikh would like to visit London. Again the spectre of American reaction haunted Whitehall. If the Shaikh came after APOC had been freely awarded the concession, things would be made a good deal easier diplomatically in the official view.

Arrangements were made to receive the Shaikh with every honour and courtesy short of a state reception of the kind which would intensify American suspicion. It was agreed that he should be asked to arrive not later than 20 July, after which date King George V would be out of London. Shaikh Ahmad had, decided to travel privately at his own expense.

The bargainers were back where they started, or worse, with Chisholm and Dickson waiting in the wings, Holmes in disgrace and Shaikh Ahmad immersed in a very delicate exchange with Ibn Saud. Nothing in this affair was uncomplicated. No sooner had the ruler decided to abandon his proposed visit to London than King George decided he would like to receive the Shaikh at Buckingham Palace. The officials of the Foreign Office had to dance with great delicacy. The British government and its representatives in the Gulf meanwhile continued their guessing game as to the intentions of Ibn Saud, Shaikh Ahmad and Holmes.

Then in July, Lt Col Fowle reported: "I asked the Shaikh how he felt about an amalgamation between the rival parties, supposing they decided among themselves at any stage that this would be more economical than the continued bidding against each other. He seemed to be confident . . . he could manage affairs to his own advantage." For the next three months the negotiations remained in suspense. In November it was reported that Holmes, who had been in England, was returning to Kuwait and that Eastern and General Syndicate had formed a new oil company in London, consisting of an American and an English group. In fact, Holmes made only two very transitory visits to Kuwait between the end of May 1933 and February 1934, when he was to return in another guise. In August, the Board of Trade wondered whether it would be possible to "get further light" on a rumoured arrangement between APOC and American interests. The India Office, egged on by the Admiralty, began to wonder whether Britain should revert to an insistence on exclusivity. They contented themselves with the suggestion that HM Government should be consulted before any new negotiations were undertaken. On the negotiating front all was quiet during almost the entire second half of 1933. It was only in December that things began to happen again.

Mr W. Fraser, deputy chairman of the Anglo-Persian Oil Company, arranged to inform the Foreign Office of his "position and difficulties" on Monday 11 December. To the surprise of the government representatives present he explained that his company had made an arrangement (extending to exploitation) with American interests, and that it had done so with the specific approval of HM Government. He said that he made that statement on the instruction of Sir John Cadman and, when pressed, intimated that he was ready to give further details in writing. On 13 December, the India Office wrote to Mr Fraser, informing him that the department concerned had been unable to trace any record of the prior approval of HM Government of an arrangement between APOC and the Gulf Oil Company. A week later the Foreign Office was informed that Mr Fraser had made his statement at the insistence of Sir John Cadman, and that Sir John wished to explain the position in person.

The deed was done, whatever HM Government might think about it. Faced with a *fait accompli* the Admiralty insisted that local refining should be in British hands and independent of any agreement with the Americans. It also wanted a British right of pre-emption in the event of war. APOC refused adamantly to have anything to do with such an arrangement and said that it would surrender its part in the agreement rather than insist on the Admiralty's provisions. On 4 January 1934, Sir John Cadman and Mr Fraser called at the India

Office to discuss the British government's attitude to the fifty-fifty arrangement and to such matters as the employment of British personnel in Kuwait, local refining and a wartime pre-emption clause, the procedure to be adopted in further negotiations with the Shaikh and the Gulf Oil Company's representation.

Sir John explained that from previous discussions with the Foreign Office and the Petroleum Department he had gained sufficient reason to believe that there would be no objection on the part of HM Government to the fifty-fifty arrangement with Gulf Oil. Not quite the "evidence" promised by Mr Fraser but the government seemed satisfied with the explanation. It was decided that future negotiations should be conducted by Chisholm and Holmes acting together on behalf of a jointly owned Kuwait Oil Company, and that they should keep the Political Agent informed. Sir John promised that should Major Holmes show any sign of causing trouble he would be withdrawn. Better take Holmes along with them, despite his capacity for mischief, than risk having him against them, they decided.

At a subsequent meeting the Gulf representative objected to the proposed insistence on British personnel in working the concession. He insisted on a qualification which would leave the company discretion in the matter. Two agreements were to be drawn up, one for the actual concession, the other between HM Government and the new company, designed to protect British interests in the event of war.

By 12 January, the India Office had agreed a draft concession prepared by the oil companies for consideration by the Foreign Office. This was finalised at a meeting attended by Mr Fraser and Mr Hearn of APOC and Mr G. Stevens of Gulf, and presided over by Sir Louis Kershaw, at the India Office on 25 January. The Kuwait Oil Company was registered in London with a capital of 50,000 one pound shares, divided equally into "A" and "B" units on 2 February 1934, with the Anglo-Persian Oil Company and the Eastern Gulf Oil Company having equal shareholdings. By 6 February, the draft concession agreement bearing its name was ready for transmission to Kuwait. News of an oil strike at No. 4 well in Bahrain had come through by this time, thus lending a renewed air of urgency to the Kuwait negotiations. The other agreement, between KOC and the British government, had also been drafted. It retained the provisions for registration within the British Empire and for the pre-emption of all crude oil supplies in time of war. Mr Stevens wanted to strengthen the wording relating to the equal shareholdings of APOC and Gulf, but the Admiralty made it clear that in its view nothing could be done to prevent commercial concerns from altering such arrangements if they so wished by "all kinds of cunning devices".

It may have been coincidental that the Treasury chose this moment to review the financial and voting arrangements governing the relationship between HM Government and the Anglo-Persian Oil Company. In doing so it insisted that an ex-officio government director of the company would have the power of veto. This would be used only in matters of foreign, military and naval policy, *and in matters affecting the control of new exploitation sites.* On 16 February the shareholding provision of the government/Kuwait Oil Company came under discussion again. The relevant clause read: "The Kuwait Oil Company shall remain a company in which not more than 50 per cent of the capital and voting rights shall be directly or indirectly controlled by persons other than British subjects." It was now agreed that this could be omitted or modified provided APOC undertook to give the government prior notice of any proposed change in its holding or voting powers. Sir John Simon approved the draft agreement between the government and the company on 27 February.

On 15 February, Chisholm and Holmes returned to Kuwait and negotiation with Shaikh Ahmad was resumed. The ruler had been informed of the agreement between the other parties and of the formation of the Kuwait Oil Company. It was, for him, a novel experience to see Holmes and Chisholm on the same side of the table. The discussions proceeded until June. Meanwhile, several small dramas developed alongside the main dialogue. The Shaikh had been extremely unhappy about the possible interaction of the two agreements involved, the one between the company and the British government, the other between the company and himself. As far as HM Government was concerned, the latter was subsidiary to the former in certain vital respects. In a nutshell, it was held that Britain should have the last word with regard to the appointment of a chief local representative of the oil company, that political considerations should be paramount in any cancellation of the agreement, and that an arbitration procedure approved by the government should decide any disputed questions. Thus, the Shaikh felt that his access to information on all fronts, including exploration, production figures and royalties could be restricted and that Britain could, if the oil company failed to comply with the terms of the political agreement, use its powers to counteract the commercial undertakings. Bargaining over two vital clauses caused a rift not only between the Shaikh and the negotiators, but between the negotiators and their principals in London. By April two new clauses had been drafted, but they did not please Shaikh Ahmad. While the drafting went on, the first rumour of outside intervention came to the surface. At an India Office meeting on 17 April, there was reference to an intelligence report which showed that the Shaikh had received a counter offer from a substantial and totally British company. He had, in fact, told Holmes and Chisholm of this offer on 11 April. Outwardly, however, the KOC negotiation was proceeding with no more than the familiar hurdles to be jumped. On 24 April, Dickson reported a meeting with the Shaikh who was at camp with his family in the desert. He was, said the Political Agent, nothing like as unsympathetic to the content of the political agreement and the clauses in dispute as was imagined. Arab pride had been affronted, said Dickson, probably because neither Chisholm nor Holmes spoke Arabic and had to rely on interpreters. He believed that the ruler felt keenly the lack of consideration shown by the manner in which the offending articles had been presented to him. Nonetheless, on 2 May, the negotiators received a letter from Shaikh Ahmad which read: "I have informed you of the terms and additions which I require in the concession and the amount of the payments and royalties, which are final. . . . I forward this communication so that you may consult your company urgently, and inform me of your company's opinion as I see no good in delaying."

The following day, the office of the Secretary of State for India informed the Political Resident: "In reference to other possible British competition, the Shaikh appears to be either bluffing or to be misinformed. We know of no other purely, or 50 per cent, British company willing and able to give the terms offered by KOC."

The Resident was also told: "For your confidential information it is desirable that negotiations should be concluded as soon as possible because APOC are anxious to get them out of the way before approaching the Shaikh on behalf of IPC with a view to obtaining an option on the Kuwait Neutral Zone concession."

On 3 May, Dickson had written to Bushire expressing the concern of Chisholm and Holmes at the Shaikh's ultimatum and the threatened outside intervention. Holmes, nervous of the activities of the mysterious interloper, announced that he intended to force the issue with KOC, who were certainly acting with no apparent haste. Both he and Chisholm complained that their principals had no conception of the problems or atmosphere in Kuwait, according to Dickson. As for Shaikh Ahmad, he was increasingly convinced that he was being bullied by KOC and HM Government. "I am concerned that he may break off the talks for good," said the Political Agent. At this stage, the negotiators seem to have taken matters into their own hands. They redrafted the agreement to the satisfaction of the ruler. London promptly rejected this fourth draft, however, and announced that it was giving serious thought to sending out some of the KOC principals to continue the talks. The India Office insisted that Chisholm should not be withdrawn without prior consultation with the government. The Political Resident, looking for somebody to blame for this never-ending see-saw of dissension and distrust, turned to the universal whipping boy Holmes. In fact, he went further and blamed KOC for appointing him against the wishes of HM Government. There were even suggestions of a secret agreement between the Shaikh and Holmes.

Not without cause, Shaikh Ahmad decided to call a halt. He told Chisholm and Holmes on 2 June that failing unconditional acceptance of his terms they should leave Kuwait as he would not discuss the matter further until September. He promised that he would not negotiate with other parties in the meantime. But on instructions from London, the negotiators continued to make representations and in the end the ruler announced that they should withdraw forthwith. The two men packed their bags and made their separate ways to London.

It was within the framework of this dispute that the intervention of a third party became more serious, and perhaps more sinister. By the time the freshly-instructed negotiators arrived back on the stage, Lord Lloyd was flitting behind the scene in London. The argument about the appointment of a chief local representative, the arbitration clause, the amount and basis for assessing royalties, and the implications of the *de facto* political agreement (signed by the company and the British government on 5 March of that year), was resumed with vigour in September. Eventually, one of the main obstacles was removed by taking the offending clauses from articles 11A (d) and 20 of the agreement. In their stead, Colonel Dickson was instructed to write them into a secret letter to the Shaikh, with an undertaking that this would not be published or divulged. In retrospect it is difficult to see what the argument was about. For example, the wording of one clause, which met with the Shaikh's approval was: "One of the superior local employees of the company shall be maintained in the region of the Gulf as Chief Local Representative." The wording

KOC wanted was: "The company shall designate its General Manager or one of its other principal employees as Chief Local Representative." It was Dickson who, at the insistence of the Shaikh, got the job when the talking was over. Therein perhaps, in matters of trust and personality, lay part of the cause of the dispute.

It was, in fact, on 12 April 1934 that Chisholm and Holmes first informed KOC in London that a counter offer had been made for the concession. Yet the bargaining over the KOC draft agreement proceeded until November, with the break insisted upon by the ruler from mid June until the end of September, without outward sign of concern. By 24 October, financial terms were agreed in London. These consisted of payment to Shaikh Ahmad of Rs475,000 initially, followed by a further Rs95,000 per annum before declaration of viable mineral deposits, Rs250,000 per annum after declaration and a royalty of two rupees and 15 annas per ton. The Shaikh accepted the down payments but rejected the royalty figure, insisting instead on three rupees and four annas per ton. Then, on 19 November, the Secretary of State telegraphed the Political Resident informing him that the government had received "secret" information that the Shaikh had actually agreed to grant to another group an oil concession over his territory, subject to HM Government's approval.

At this point, Lord Lloyd emerged from the shadows. The Foreign Office received a telephone call from the noble Lord at about this time, in the course of which he informed them that he had reliable information that the Hunting group of companies, acting as Traders Ltd, had concluded an agreement with the Shaikh. He mentioned that Mr P. H. Hunting was a director of British Oil Developments which, in turn, was closely allied to Mosul Oilfields Ltd. But in this connection, said Lord Lloyd, Mr Hunting was acting independently. The communicant was at pains to make it clear that he had no vested interest in the matter, that he did not wish his name to be in any way connected with the application but that he hoped that the government would give favourable consideration to the claims of an all-British concern against those of a half-American consortium. Some urgent enquiries were put in hand. The Foreign Office knew about Lord Lloyd and his concern for Britain's imperial supply lines. Sometime Member of Parliament for West Staffordshire and later, for Eastbourne, he had a considerable knowledge of the Middle East and of naval affairs. He had been High Commissioner to Egypt and the Sudan, and HM Special Commissioner for trade in Mesopotamia and the Persian Gulf. At the time he was President of the Navy League. His representations on behalf of Mr Hunting could not be dismissed lightly. He appeared, according to the Foreign Office, to be hostile to APOC.

What of Mr Hunting? Up to 1933 he had been a director of two companies, one with a registered capital of £215,000, the other with £160,000. His principal interest was in the manufacture of hemp and wire rope. Since then, however, his directorships had expanded to include eight companies, one of which was Traders Ltd, registered at Newcastle-upon-Tyne. Traders had an impeccable board. Apart from Mr Hunting, its members were Lord Glenconner, Mr Edward Tennant of Charles Tennant Sons & Company Ltd, Sir Richard Redmayne and Mr H. H. Holmes (no relation to Major Frank Holmes).

On 21 November the Political Resident in the Gulf telegraphed the

Secretary of State for India that the Political Agent had communicated some vitally important information to him and that he would, after instituting some further investigation, contact London on the following day. The PA's information turned out to be that in August, while he, Dickson, was on leave and local Kuwait affairs were in the hands of Major Ralph P. Watts, the Shaikh had been approached by an Iraqi with a draft concession which improved on the terms offered by KOC. He suspected that the company involved was Frank Strick & Co. of London and Basra, but "presumably there is another company behind them", he suggested, with obvious reference to Traders Ltd. The India Office, on receipt of this information, told the Foreign Office: "The Shaikh appears to have behaved quite properly in the matter, having told the interested party that any agreement is subject to HM Government approval."

The head scratching that went on at this juncture gave rise to at least one interesting recollection. In April, while the oil talks were at their peak, the German Minister in Baghdad, Dr Fritz Grobba, together with the French Chargé d'Affaires, M. Lepissier, had paid "a hurried and not particularly welcome call", to use the Political Agent's own words, to Kuwait. They were accompanied by a large retinue and asked questions which seemed more than pointed about the prospect of finding oil in the territory.

To return to the situation prevailing in November 1934, more of Col Dickson's communication to the Resident began to come to light following the latter's insistence that some of the PA's allegations should be checked. In short, Dickson had spoken to the Shaikh who was said to have affirmed that the approach came from Frank Strick & Co. working through a Basra lawyer K. B. Mirza Mohamad, "in secret partnership with the lawyer Gabriel". He went on: "The following facts are significant. F. Strick & Co. are connected with APOC and receive an annual sum from them, Mirza Mohamad is a legal adviser to APOC and is also lawyer to F. Strick & Co. Mirza Mohamad and Gabriel are partners in the same firm." He went on:

'Is it beyond the bounds of possibility that APOC, smarting under the fact that they were unable themselves to get an oil concession out of the Shaikh because of the Americans, and while outwardly agreeing to work with them under the name KOC, secretly set themselves the task of hindering progress from the start with the deliberate intention of preventing the Shaikh from giving the concession to the Kuwait Oil Company, and at the psychological moment and when the negotiations with KOC had reached deadlock they had ready a new and secret subsidiary company to throw into the breach, and so walk off with the prize?'

Whatever suspicions Dickson's theories may have given rise to in government circles, the Foreign and India Offices exhibited their usual calm front.

Mr Stevens, the Gulf nominee on the KOC board, complained at a meeting with government officials a few days after the Political Agent's communication had been received that his company had not been informed as quickly as he might have hoped about the Traders' intervention. He asked for final government approval of the draft between KOC and the Shaikh so that he, Stevens, could return home

IN THE NAME OF GOD THE MERCIFUL THE COMPASSIONATE.

AGREEMENT made on this day of one thousand nine hundred and thirty four, corresponding to one thousand three hundred and fifty three Hajri, BETWEEN HIS EXCELLENCY SHAIKH SIR AHMAD AL-JABIR AS-SUBAH, Knight Commander of the Most Eminent Order of the Indian Empire and Companion of the Most Exalted Order of the Star of India, the Prince and independent ruler of the State of KUWAIT (hereinafter called the Prince) and TRADERS LIMITED of Milburn House in the City and County of Newcastle upon Tyne England, a Company of British invested capital registered in Great Britain under the Companies Act 1929, its successors and assigns (hereinafter called the Company) WHEREBY the Prince grants and the Company accepts the Kuwait Oil Concession as hereinafter described under the following terms and conditions

ARTICLE 1.

SUBJECT to the sanction of His Britannic Majesty's Government the Prince hereby grants to the Company THE exclusive rights to explore, prospect, search and drill for, produce, extract and render suitable for trade, Asphalt, Ozokerite, Petroleum and their products, derivatives and cognate substances (hereinafter terms "Oil") and also naturual gas and its products and derivatives, and the rights to carry away and sell the same within the State of Kuwait including all islands and territorial waters appertaining to Kuwait as shown generally on the map annexed hereto, together with the exclusive ownership of all oil and natural gas produced and won by the Company within the State of Kuwait for a period of SEVENTYFIVE YEARS beginning from the date of

37. The agreement that never was, with Traders Ltd

to America for Christmas. If Dickson was right and APOC was finessing, it was surely playing a dangerous game since a Gulf representative was in regular communication with the British government. However, nothing of the British suspicion in the matter was ever mentioned at meetings attended by Mr Stevens. The India Office was leaving no stone face upward. It asked Sir Henry Strakosch, a man of great knowledge and renown in city matters, for a reference for Mr Hunting and Traders Ltd, with particular regard to their probity, technical qualifications and resources. They also asked the Department of Overseas Trade to look into their *bona fides*. Strakosch told Sir Cecil Kisch, Under Secretary of State at the India Office, in a confidential letter: "Judged by its capital the most important enterprise with which the group is connected is Mosul Oilfields, housed in the offices of Charles Tennant Sons & Co." Of sixteen directors, Sir Henry Strakosch thought that eight, and possibly nine, were foreign, mostly Italian. Mosul's original capital was £1 million. On 6 April 1934 that capital was increased by the issue of one million "A" shares with a nominal value of 6s 8d each. In September there was a further infusion of cash obtained by issuing another 500,000 "A" shares of the same face value. Generale National Petroli of Rome and Deutsche Teuhand Gesellschaft of Berlin took up the majority of these shares, according to Sir Henry, who calculated that at most the British interest was 51 per cent.

Thus informed, the British government began to take a firm line. On 4 December, Traders Ltd officially requested government approval of its application and on the same day Lord Lloyd renewed his telephone intercession on behalf of the applicants. He threatened to raise the matter in the House of Lords if a totally British owned company was excluded from the concession. He repeated, however, that he had no personal interest in the matter and conveyed with some anxiety a wish that nothing he had said should be so interpreted as to embarrass Shaikh Ahmad. By 12 December, Lord Lloyd had modified his approach. Could he take it, he asked, that if he did not raise the matter in Parliament, the Foreign Office would hang fire on the KOC agreement? On 16 December, Dickson saw the ruler who explained that in September he had permitted the lawyer Gabriel to show him the draft terms of Traders Ltd. He had expressed approval of the terms, subject to Britain's acceptance. He had heard no more, however, and on the previous day, 15 December, he told them that he had reached agreement with the KOC negotiators and that he could receive no further approaches from them.

The ruler's qualified acceptance of the Traders proposal, sent to the lawyer Gabriel on 2 September 1934, concluded:

> 'I . . . return a copy (of the draft) to you so that you may despatch it to His Majesty King George's Government in London. Should HM Government agree, I myself accept the terms.'

The KOC talks had reached the stage of a fifth and final draft. There can be no doubt that the negotiations were accelerated by all three sides in their later stages. The agreement was finally signed on 23 December 1934. The day before, the Admiralty had signalled its approval to the Foreign Office which, in turn, gave its sanction to the Resident.

Colonel Dickson telegraphed: "I have made the necessary communication to the Shaikh who signed the concession in my presence today (23 December)."

The agreement as amended gave an exclusive right to the Kuwait Oil Company to explore, search and win natural gas and crude oil (as shown on an annexed map), as well as the rights to refine, transport and sell those products within and without the state. The company contracted to begin exploration within nine months and aggregate depths of drilling were specified as and from certain anniversary dates after signature, with a maximum depth of 30,000 feet prior to the twentieth anniversary. The company would have the right to maintain in the region of the Gulf a chief local representative, but in the first instance the Shaikh would have the right to select this representative in consultation with HM Government. The Shaikh could appoint an Arab official to represent him locally and a Kuwait, British or American subject as his official representative in London. The Shaikh was to provide trustworthy guards to protect the company's property. At any time after drilling to 4,000 feet the company could give one year's notice of its intention to quit. In the event of the company not fulfilling certain agreed obligations, within specified time limits, all the company's properties within the state would become the Shaikh's.

The Amir was paid £35,625 on signing the agreement. If and when oil was found, the state would be paid three rupees per English ton (of 2,240 lbs) of net crude petroleum won and saved, plus four annas in lieu of duty, making a total of about 25p. a ton at the exchange rate of 1934. The state also received a rental of £7,125 sterling a year. The agreement was to last for seventy-five years and on expiry the company was bound to surrender all movable and immovable properties to the state.

In the meantime, on 14 December 1934, the two contracting companies made a separate agreement limiting competition between them. It was an important document, for it formed the basis of a commercial relationship that was to prove remarkably successful, though it doubtless had its marital difficulties. Its provisions were summarised in this explanation of its main clauses:

"The parties have in mind that it might from time to time suit both parties for Anglo-Persian to supply Gulf's requirements from Persia and/or Iraq in lieu of Gulf requiring the company to produce oil or additional oil in Kuwait. Provided Anglo-Persian is in position conveniently to furnish such alternative supply, of which Anglo-Persian shall be the sole judge, it will supply Gulf from such other sources with any quantity of crude thus required by Gulf provided the quantity demanded does not exceed the quantity which in the absence of such alternative supply Gulf might have required the company to produce in Kuwait – at a price and on conditions to be discussed and settled by mutual agreement from time to time as may be necessary – such f.o.b., however, not to be more than the cost to Gulf of oil having a similar quantity produced in and put f.o.b. Kuwait."

Thus ended what must surely have been the most complex and protracted negotiation for an oil concession ever conducted. Even to the last it was a near thing. From the first tentative moves by the syndicate and APOC in the early 1920s, it had taken over ten years to

II.—Commercial Agreement between the Sheikh of Koweit and the Kuwait Oil Company, dated 23rd December 1934.

IN THE NAME OF GOD THE MERCIFUL.

THIS IS AN AGREEMENT made at Kuwait on the 23rd day of December in the year 1934 corresponding to 16th day of Ramadham 1353 between His Excellency Shaikh Sir Ahmed al-Jabir as-Subah, Knight Commander of the Most Eminent Order of the Indian Empire and Companion of the Most Exalted Order of the Star of India, the SHAIKH OF KUWAIT in the exercise of his powers as Ruler of Kuwait on his own behalf and in the name of and on behalf of his heirs and successors in whom is or shall be vested for the time being the responsibility for the control and government of the State of Kuwait (hereinafter called "the Shaikh") and the KUWAIT OIL COMPANY LIMITED a Company registered in Great Britain under the Companies Act, 1929, its successors and assigns (hereinafter called "the Company").

(Signed) ____________________
(Signature of Shaikh of Kuwait.)

(Signed) H. R. P. DICKSON, Lt.-Col.

Article 1.—The Shaikh hereby grants to the Company the exclusive right to explore search drill for produce and win natural gas asphalt ozokerite crude petroleum and their products and cognate substances (hereinafter referred to as "petroleum") within the State of Kuwait including all islands and territorial waters appertaining to Kuwait as shown generally on the map annexed hereto,* the exclusive ownership of all petroleum produced and won by the Company within the State of Kuwait the right to refine transport sell for use within the State of Kuwait or for export and export or otherwise deal with or dispose of any and all such petroleum and the right to do all things necessary for the purposes of those operations. The Company undertakes however that it will not carry on any of its operations within areas occupied by or devoted to the purposes of mosques sacred buildings or graveyards or carry on any of its operations except the sale of petroleum housing of staff and employees and administrative work within the present town wall of Kuwait.

The period of this Agreement shall be 75 years from the date of signature.

Article 2.—(A) Within nine months from the date of signature of this Agreement the Company shall commence geological exploration.

(B) The Company shall drill for petroleum to the following total aggregate depths and within the following periods of time at such and so many places as the Company may decide:—

(i) 4,000 feet prior to the 4th anniversary of the date of signature of this Agreement.
(ii) 12,000 feet prior to the 10th anniversary of the date of signature of this Agreement.
(iii) 30,000 feet prior to the 20th anniversary of the date of signature of this Agreement.

(C) The Company shall conduct its operations in a workmanlike manner and by appropriate scientific methods and shall take all reasonable measures to prevent the ingress of water to any petroleum-bearing strata and shall duly close any unproductive holes drilled by it and subsequently abandoned. The Company shall keep the Shaikh and his London Representative informed generally as to the progress and result of its drilling operations but such information shall be treated as confidential.

* Not reproduced.

38. Final draft of the successful concession agreement between the Shaikh of Kuwait and KOC. By courtesy the India Office Library reach agreement. The bargaining had involved the British cabinet and all the principal British departments of state; it had reduced ministers and senior state officials to bewilderment and near apoplexy.

Had the various parties known at the time that they were competing for what was probably, acre for acre, the richest single prize in history, the battle would doubtless have been fought with even more tenacity. As it was, the issues were primarily those of Britain's imperial prestige and America's open door policy, complicated beyond any usual degree by the coincidence of a personal involvement on the part of the US Ambassador in London, by the delicate state of Kuwait's relationship with her territorially vast neighbour, Saudi Arabia, and by the other oil negotiations proceeding in and around the Gulf region at that time. It is, of course, easy to look back with wisdom. To do the right thing consistently, in the heat of battle is another matter. Yet, looking back, it is hard to resist the conclusion that all the parties behaved with predictable concern for their interests, whether of company or country.

Shaikh Ahmad kept strictly to the letter of his country's agreement with Britain, making it clear to everyone concerned that no bargain would be struck without the approval of HM Government. He was, as the British several times conceded, a sovereign ruler and his main concern was to obtain the best possible arrangement from his country's point of view. Despite inferences to the contrary, there is no evidence to suggest that anything of substance was concealed from the British. In the absence of any other serious bidder, he could hardly be blamed for negotiating with the syndicate which was, in any case, a British company, represented by a British subject, though acting for an American concern. Yet the Traders episode continued to intrigue and aggravate the participants even after the signing of the agreement. Col Dickson believed to the end that there was a conspiracy. Mr Chisholm, on the other hand, was convinced that the application was a straightforward offer by Hunting and his colleagues, advised by Gabriel who, as lawyer to the Shaikh and Traders, was in a very favourable position to help. The only official subsequent word on the subject was pronounced by Sir John Simon. Writing to Sir Andrew Ryan at Jedda on 10 January 1935 he remarked: "HMG learned by a circuitous route of Traders' interest." BOD (Mosul Oilfields) was not involved but several of its directors were, he said. There were further references to the fact that Shaikh Ahmad had been reminded on 23 April 1934 of his obligation, under the agreement of 27 October 1912, to inform HMG before negotiating with anyone. Tempers continued to boil and Dickson was still reporting – and conjecturing – well into 1935. The British government decided, in April of that year to make a further representation to the Shaikh. Dickson, against his own advice and wish, delivered the official reprimand on 11 April. In it the ruler was asked for an assurance that no such negotiations would be entered into again without prior approval from HMG. It brought forth a short and pointed reply, dated 21 April. Its final paragraph read:

"Seeing therefore that we are independent, under the shade of His Majesty's Government's protection and good will, and seeing that we are holding tight to everything that helps to confirm this fact in every way, we cannot agree to the above mentioned directions, except insofar as they concern oil concessions in the neutral zone, for we believe that we have absolute liberty in our hand."

As for the Anglo-Persian Oil Company, it could certainly be argued that it would have saved the British government a great deal of time and temper, if it had kept it more closely informed. If there was any truth in Dickson's allegations, Traders Ltd, acted their part to the end. Mr Hunting took an angry view of the summary rejection of his offer, threatening to take legal action against both the British government and Shaikh Ahmad, though he discovered that it is not easy to sue the Crown or a sovereign ruler.

What of Major Frank Holmes? Vilified and suspected by everyone, except perhaps by Shaikh Ahmad who genuinely liked him, Holmes was very much the butt of the British, damnably clever when he appeared to be succeeding, a scoundrel to be kept within eye-shot when things went against him. He was certainly no committee man. He was not given to spending time and effort in administrative niceties. Neither did he try to cover up his aim of securing the concession and making a hard-earned profit in the process. The charges of deception and deviousness levelled against him by Whitehall, APOC, the Political Agent and the Political Residents, held little water. He may not have been "their kind of man", but in company that was anything but forthcoming, he was often overt to the point of indiscretion. Colonel Dickson – rigid, incorruptible, quick to take offence and a stickler for protocol – and Holmes, extrovert and cavalier in manner, were hardly the men to do business together. All the same, the Dickson family and Chisholm found the man agreeable outside of the bargaining round. All enjoyed his rollicking ebullience, off duty.

An interesting question remains. Why, if the methods of Holmes were as dubious as most of the participants maintained, did he provide the Political Agent with a detailed account of his extensive observations, complete with maps which showed a remarkable prescience if nothing else, in 1931, at the very moment when British interest in the concession was being revived? Why was the document, on the strength of which Colonel Dickson was able to draw up a clear account of the syndicate's intentions, so studiously ignored? Or did Anglo-Persian, despite Sir John Cadman's assertion to the contrary, know all the time the extent and whereabouts of the oil deposits in Kuwait?

There was a postscript to those events which, looking back, was not out of character. Shaikh Ahmad finally made his visit to London, in 1935. And he took with him the largest pearl ever found in Kuwait. It was about the size of a glass marble, perfectly round and flawless. The Shaikh intended it as a present for Queen Mary. The pearl travelled to London in the mouth of the Shaikh's personal bodyguard, the tall, powerfully-built negro Mirjan. However, Mirjan had to eat and one morning he removed the pearl from his mouth in order to consume his breakfast, leaving it on the dressing-table of his hotel bedroom. When he went to replace it in his mouth it had disappeared. Luckily the chambermaid sweeping under the bed had picked it up. She produced it as soon as the agitated Kuwaitis spread the news that it was missing, disarmingly saying that she thought it was a bead of no value.

– 15 –

Article III. — This Agreement is written in English and translated into Arabic. If there should at any time be disagreement as to the meaning or interpretation of any clause in this Agreement the English text shall prevail.

In witness whereof the parties to this Agreement have set their hands the day and year first above written.

ON BEHALF OF THE KUWAIT OIL COMPANY LIMITED

Frank Holmes

SHAIKH OF KUWAIT

IN THE PRESENCE OF

23.12.34

Lieut.-Colonel, H. B. M's Political Agent, Kuwait.

IN THE PRESENCE OF

23.12.34

Lieut.-Colonel, H. B. M's Political Agent, Kuwait.

39. The oil agreement signed by Shaikh Ahmad, A. H. T. Chisholm, Frank Holmes and H. R. P. Dickson, at the Political Agency on 23 December 1934

7. Harvest of the Desert

Several investigations of Kuwait's surface deposits had been made before the arrival of the combined British and American drilling teams early in 1935. S. L. James described seepages at Bahrah and Burgan in 1914, and in 1917 he and G. W. Halse conducted a search. In 1926, B. K. N. Wylie and A. G. H. Mayhew carried out further investigations, and they were followed by T. Kewhurst in 1931. In 1929, Major Frank Holmes drilled to a depth of 705 feet in the north-east but his well bottomed in limestone. There was also the hole by the seafront. But it was the visit of the Anglo-Persian geologist P. T. Cox in 1931–2 which produced the first authentic evidence of oil. Following this visit he reported three oil or gas seepages, and a surface anticline at Bahrah associated with a seepage. At that time, Anglo-Persian drilled two statigraphic holes, one at Bahrah and the other at Burgan. In 1935, soon after the signing of the concession agreement, a full-scale geological party arrived, led by Cox and the American Ralph O. Rhoades who later became chairman of the Gulf Oil Corporation. It recommended that an exploratory well be drilled at Bahrah.

Meanwhile, the pioneers of the Kuwait Oil Company, some drawn from Gulf and others from Anglo-Iranian (as the company retitled itself in 1935), appeared on the scene. Mr L. D. Scott, known throughout his career in Persia and Kuwait as "Snib", was in charge of the party as the first of KOC's managers. He arrived on 30 December 1935 aboard a British India mail steamer, accompanied by his wife Joan, Donald A. Campbell, an accountant who subsequently became Company Secretary of KOC, F. J. Shields, a civil engineer with Anglo-Iranian, two Indian clerks, three dogs and some twenty lavatory pans. The advance guard was shortly followed by the Americans Tom Patrick and Ted Rakestraw, and five other Britons, Danny Robinson, Peter Walker, A. R. G. Harvey, Norman Allison and Scott's personal assistant W. A. Chase.

The first Kuwaiti to join the company's payroll was Abdul Salem Shaebb. He was the nominee of Shaikh Ahmad's private secretary Mulla Saleh and his appointment, though in a junior capacity, throws an interesting light on the wary relationship which existed between the Amir and the oil company even after the ratification of the agreement. According to Shaebb himself, he was lying in his tent along the Bay of Kuwait recovering from pneumonia when Mulla Saleh's messenger called him. He was told to dress and repair to his home in Kuwait town where he would be instructed on the requirements of his job as clerk to the company. In fact, he was told to report regularly to the Shaikh's secretary and to inform him of every detail of the exploration without delay. A salary of 80 rupees a month was negotiated with the company. He was also promised payments, which in the event varied from 200 to 300 rupees a month, for his extra-mural labours.

Soon after the arrival of Scott's contingent, KOC's first office was opened under the direction of Donald Campbell. He and Abdul Salem Shaebb were quickly joined in competition for its limited facilities by the firm's second Kuwaiti employee, Issa Abdul Jelil. But administrative refinements were of preciously small concern to L. D. Scott's team, whose drillers were anxious to get on with the practicalities.

The Cox and Rhoades report, dated 1 June 1935, had recommended:

1. That a well capable of reaching 6,000 feet be drilled on the Bahrah anticline at a location on the raised ground near the gas seepages. This well should be cored continuously and should be under the observation of a geologist at all times.
2. Simultaneous with the drilling of the well at Bahrah a geophysical survey should be undertaken in the area south of Kuwait Bay, to be extended from the coast to the Shaqq depression and the Jal es Zor. Particular attention should be given to the area along the line through Madaniyat and Burgan. Preliminary to this survey a few profiles should be placed across the Bahrah anticline.
3. Following the completion of the well at Bahrah and regardless of its results or of the results of the geophysical survey, wells should be drilled at Madaniyat and Burgan. Depths to which these wells are to be drilled may be influenced by information obtained at Bahrah.
4. Topographic mapping as recommended in the area mentioned in recommendation 2.

The drilling rig was assembled at Bahrah in May 1936 and the first wildcat well was spudded on 30th of that month. It was a difficult operation, beset by all the problems of remoteness and lack of adequate workshop and supply resources. To break a piece of equipment at that time was a disaster according to L. D. Scott, who often waited impatiently for weeks on end for a replacement to arrive from Persia. They worked for almost a year – until April 1937 – encountering modest signs of oil but none of them suggestive of commercial quantities. They penetrated a total depth of 7,950 feet before they finally acceded to the drillers' dogma that oil is not so much where the geologists say it ought to be, as where you find it.

Further geological surveys were carried out by Cox, Rhoades and another American Paul Boots, between 1936 and 1937, using gravity, magnetic and seismic methods. These revealed the likelihood of a structural formation near the Burgan bitumen deposits. Thus, the team dismantled derrick and drilling equipment and moved it to Burgan, where it was reassembled in October 1937. This was no easy task in the blazing heat of the desert. Those who speak of delays in the early exploration of Kuwait's oilfields are, perhaps, misled by the simplicity of modern drilling operations in which vast rigs are hauled from one place to another on giant bogey trailers in a matter of hours. Then, the task was piecemeal, and backbreaking in its rigours.

The scientific reasons for moving from Bahrah at the north of Kuwait Bay to Burgan in the south, were reinforced at the time – at least in the minds of the Arabs of the desert who watched the operation with some interest – by a dream of Colonel Dickson's which has become legendary in the story of Kuwait's oil find. Dickson, who recorded his dreams each day in his diary, relinquished the post of Political Agent on 4 February 1936 and, at the request of Shaikh Ahmad, joined the Kuwait Oil Company as chief local representative. He tells the story of his premonition in *The Arab of the Desert*.

"My wife and I were living in a large oil camp in a small bungalow with large compound. In the centre of the compound, on the south and rear side, was a large sidr tree – the bungalow looked out on a sandy waste.

"One night a great windstorm arose and blew with a violence I had never before experienced. All through the night the gale blew and windows and doors rattled dismally, whilst the smell of the dust which came down with the storm penetrated thick into the interior of the house and made breathing difficult. It was a never-to-be-forgotten night. When dawn broke my wife and I looked out and saw that the storm had abated, and strangest of all saw that the wind had excavated a great cavity in the ground near the sidr tree at the back of the bungalow, about 100 feet by 100 feet and 6 feet deep. Isolated and standing in the centre of the cavity was a sort of masonry table, surmounted by a large stone slab six feet by four feet in size. On this platform lay a prone figure, shrouded in an ancient yellow-coloured cotton cloth. My wife and I went out and examined the figure and removed with care the covering from the face and head. Age had rotted the cloth and portions kept coming away in our hands. To our

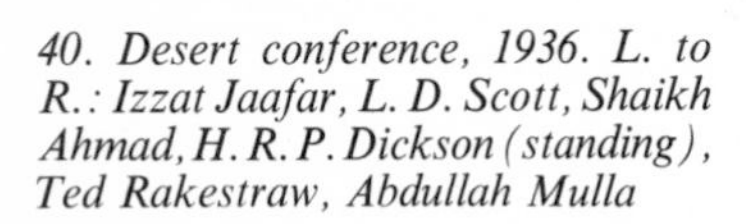
40. Desert conference, 1936. L. to R.: Izzat Jaafar, L. D. Scott, Shaikh Ahmad, H. R. P. Dickson (standing), Ted Rakestraw, Abdullah Mulla

surprise we saw that the face was that of a young and once beautiful woman, with skin hard and mummified, and in colour parchment brown. Clearly the storm had unearthed an ancient tomb, and had blown away the surrounding sand so as to leave the occupant lying on her original stone bed, now elevated like a table. It was a gruesome find and I hurried in to find some servants to help me dig a fresh grave. When I came out again my wife in great excitement pointed to the face of the dead woman, which showed distinct signs of undergoing a change – it was coming alive. As we watched in horror-stricken amazement, the parchment skin grew soft, colour came into the dead cheeks, and the body began to breathe. Then the eyes opened and the woman sat up and began slowly to disentangle herself from her shroud. Turning her head the woman saw us, and said in what must have been ancient Arabic, 'I am cold after my thousands of years' sleep – help me with some warm clothes and give me something to eat.' She then handed us a very ancient copper coin. We took her by the hand and led her into the house, where our Arab maid washed and dressed her while we prepared food.

41. Night view of Bahrah rig

"After partaking of a meal the woman expressed a desire to go into the garden and sit under the sidr tree. This she was allowed to do. Presently she looked up and said. 'I am in great danger, and must go to the British Consul or to His Highness the Shaikh. They only can protect me.' When asked to explain she said, 'Certain wicked men if they knew that I had come to life will seek to kill me and bury me again. They cannot bear me to see the light of day. Save me from them, friends, for I want to live.' Even as she spoke we heard the shouts of a crowd of men and saw bearing down on us many persons carrying staves and swords obviously intent on harm. They were led by an old white-bearded man who looked like a Persian. In his hand was a long knife, and his face was very wicked to look on. Coming closer he shouted to his satellites to seize the now shrinking form of the woman and prepare to bury her alive. Whilst some bound her, others began frantically digging a deep grave, while a third party filled the grave with water. My wife and I seemed to become suddenly paralysed and were quite unable to move hand or foot to help the girl.

"'We shall bury her alive and in a grave of mud,' screamed the white-bearded ancient in a frenzy, and assisted by others dragged the girl to the graveside till her legs rested in the mud grave hanging over the edge. As she struggled I felt the spell break, and was able to jump forward and rush to the girl's assistance. Killing the old man with a blow on the head, I attacked the others in berserk fashion and in a few moments no one was left but the old man, my wife and the girl. Taking the poor girl by the hand and putting a cloak around her we took her into the bungalow. Then I woke."

Dickson went on to explain how he had told an old woman of the desert, Um Mubarak, of his dream, and asked her what it meant. She told him to go to the chief of the drillers, Mr Scott, and advise him to seek oil by the sidr tree.

Oil technicians are conservative people. Unlike prospectors and financial speculators, in whom hope seems to spring eternal, they expect what they find.

They drilled for nearly five months at Burgan, returning to their homes in Kuwait late each night, having a few hours rest and going back to work at dawn the next morning. It was 1938 and international events were beginning to impose a sense of urgency on their activities. On 22 February, they struck oil-bearing sands and this "top show" was enough to convince L. D. Scott that they were not far from a productive zone. On the 25 February, H. T. Montague Bell, a correspondent of *The Times* newspaper who had been staying with the Scotts in Kuwait, was told that they had found oil at Burgan, but Scott was unable to tell him the extent of the discovery. Montague Bell was asked to say – or at any rate to write – nothing for a few days. Meanwhile the drillers returned to work.

The first real flow of oil came in the early hours of 26 February. According to Scott, not a man to exaggerate, it was a "difficult well" in its early stages. The story of what followed is best told in his words.

42. The new technology. Road graders in action, watched by Shaikh Ahmad, 1937

"On the night when the main oil horizon was struck, at the inconvenient hour of 1 a.m., Tom Patrick, the drilling superintendent, and the duty driller took exactly the correct action. They pulled the drilling tools up into the cased hole and continued to circulate mud fluid in order to keep the well under control and prevent the drill pipe from getting stuck. Fresh mud was pumped into the casing but it soon became evident that mud stocks were becoming exhaused. We then decided to plug the hole with a ship's mast. Donald Campbell obtained a suitable pole from one of the shipyards which we tapered to a length of about ten feet. Then, the drill pipe was pulled out of the hole and the mast driven by the pipe into the pilot hole, giving us a complete closure of the formation. The main valve was then replaced."

This down-to-earth explanation of a night of frantic activity needs elaboration. To say that the great gush of oil which came up from 3,950 feet below the surface was "difficult", is perhaps to understate the matter. Before the pole was inserted, the main valve at the head of the casing had jammed under the enormous pressure of the oil coming to the surface. Efforts to keep the well under control were not helped by a shortage of mud or barytes (barium rock) needed to quell its force, or by the fact that a deluge of rain came to drench the tired drillers. The task of finding and transporting a ship's mast was no easy one. Improvisation was of a classic order. After the mast was run to earth at a shipyard it had to be sliced down the middle and chamfered to a cone shape before it could be rammed into the hole. Meanwhile an extra 250 tons of barytes was needed. This eventually came from Abadan and was rushed to the well head by transport provided by the ruler and two prominent Kuwaitis, Yusuf al Ghanim and Saleh Othman. Burgan No. 1 well was finally completed on 14 May 1938, some three months after the discovery of 26 February.

On that momentous day in Kuwait's history, the oil men were naturally keen to deliver the news to Shaikh Ahmad. It was late afternoon when they arrived back in Kuwait, wet and tired. Mr Scott went off to see the ruler, who listened with great interest and customary politeness. But, as it happened, he knew already. Earlier that afternoon the Dicksons had taken a drive and, as they approached the gateway in the wall of the old town they met Shaikh Ahmad taking his wife for a drive. As the two cars came to a halt Colonel Dickson got out to convey his respects. The Shaikh took his arm and said, "Thanks be to God. They have found it." Then he made an unprecedented demand. "You must congratulate my wife," he said. To be asked to address the Shaikh's wife, who was of course in purdah, was so unusual that only the most prodigiously important event could demand it. Oil fever is, of course, infectious – and it seems that the ruler's earlier reservations were being overcome by it. But neither the Shaikh nor the oil companies knew at this stage just how prodigious that particular event was to prove.

Montague Bell, incidentally, was as good as his word. *The Times* report of the oil discovery in Kuwait appeared on 1 March 1938.

That report drew an interesting letter from Sir Louis Dane. Writing to *The Times* of 8 March, he described how, when he had travelled with the Viceroy, Lord Curzon, to Kuwait in November 1903, they were told that a Mr Reynolds had asked to see them. Sir Louis was "rather surprised" as he had been told that there were no Europeans there. Reynolds was seeking a passage in one of the ships accompanying the Viceroy so that he could catch the mail steamer from Bushire. He said that he had heard rumours of oil in the desert near Kuwait and had gone out by camel to prospect, but he had found no traces. Curzon therefore encouraged him to go and look for oil near Ahwaz in Persia, "from whence his favourable report led to the formation of the D'Arcy syndicate and later the APOC". Such are the seeds from which industrial giants grow.

As for Colonel Dickson's dream, Kuwait's discovery well was certainly near a sidr tree, a rare enough plant in the desert.

Soon after the Burgan strike, J. A. Jameson, by now technical director of the Anglo-Iranian Oil Company, visited Kuwait together with Dr G. M. Lees, the company's chief geologist. Jameson (JJ throughout the world of oil) suggested that some of the oil captured from the discovery well at Burgan should be bottled and presented to

43. Shaikh Ahmad visits Bahrah, 1936. L. to R.: F. J. Shields, N. Allison, Ted Rakestraw, one of the ruler's aides, Shaikh Ahmad, Dr Mylrea

the ruler. Kuwait Oil's first manager L. D. Scott and Colonel Dickson presented the Shaikh with six bottles of the fluid from that historic initial flow, at the Dasman Palace in April 1938.

Two years had elapsed from the disappointment of Bahra to the final success at Burgan, four years from the signing of the agreement. Looking at the matter from a comfortable distance, some observers have suggested that such intervals were unnecessarily long. But it is easy to be wise in retrospect. For those who were on the spot, the task of erecting a rig in the great heat of the Gulf day, the hostility of the desert and its climate, were not the conditions they would have chosen had they set out to break records. At any rate four years it was, and by the time oil began to flow in Burgan it was the year of Munich.

In the next two years they drilled eight more wells. All but one, a dry hole at Madaniyat, were productive. The drillers and technicians had pitched their assortment of tents among the bedu at Magwa, north-west of Burgan. A clinic was set up there. Drugs, dressings and medical facilities generally were in short supply, but they were supplemented by the American Mission hospital in Kuwait town.

Theirs was a small community. Whether they lived in a tent or in one of the traditional mud-brick dwellings in the town, they kept open house. Water, as always in the desert, was an ever-present problem. The history of the Arab has been largely dominated by long and tortuous journeys between distant watering places and when the Europeans joined them they shared the need, and the problem. Later, Captain C. O. (Tommy) Tucker came from Abadan to take charge of marine facilities and he organised a water shuttle service between Kuwait and Shatt al Arab in Iraq. But for the time being, Kuwait endured a shortage of both water and foodstuffs. The war, with attendant political problems in Iraq and the arrival of a large number

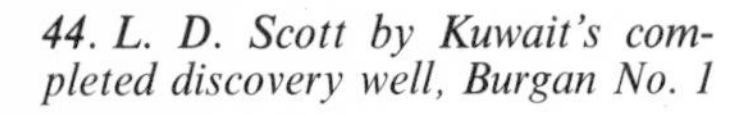

44. L. D. Scott by Kuwait's completed discovery well, Burgan No. 1

of troops in the Gulf, brought a good deal of hardship to Kuwait. Many bedu of the Kuwait and Saudi Arabian desert regions were near to starvation. As for the British and American oil men and their families, the Commander-in-Chief Middle East was making the decisions. Late in 1942, they were told to make their way back to Abadan. The well heads were plugged with concrete and Col Dickson, along with Danny Robinson, stayed to guard the oil company's nine static oil wells and several water wells which were left in operation so that the bedu could use them. The last of the Kuwait Oil Company employees left in July 1943.

Kuwait's first oil export

Before the oil drillers departed in 1942–3, another agreement was signed by Shaikh Ahmad and the Eastern Gulf Oil Company, the subsidiary formed to negotiate for Gulf in the shadow of the red line agreement. This concerned sulphur rights in the state. An optional agreement was signed on 5 June 1941 and the concession ratified on 20 November in the same year. The first of six test holes was drilled in October 1944. The results did not live up to the hopes of the signatories, however, and the concession was renounced in November 1949.

With the end of the war, the oil field development programme was resumed. The men who returned in 1945 were substantially the pioneers of 1935 to 1942, but they were soon joined by a host of American and British drillers who, during the next few years of massive expansion, changed the face of the desert.

45. Bedu guard

Local management, led by L. D. Scott and his deputy, the pleasant and able American Tom Patrick, now took up its station in Kuwait town, just by the western gateway in the old wall. They called it Marble Arch, though its official name was the Behbehani Compound.

General Pyron, a Gulf vice-president recently returned to civilian life from the Texan 4th Cavalry Division, appeared at intervals to find out how things were progressing. His approach was never less than brisk. He wanted to see oil in the pipeline without delay – by what means or method he was not greatly concerned, so long as it did not cost too much.

The small improvised workshop at Magwa was now overflowing as mechanical equipment came in from Abadan and elsewhere. Earth movers, rigs, tractors, diesel motors and compressors, automobiles of every size and description littered the landscape. The men, probably the highest paid manual workers in the world, formed their groups and factions in this desert society, instituted their own unwritten laws, and in the manner of oil drillers played out their leisure time in thriving card schools. The colourful scene was given added zest by the appearance of a group of Texan wives, and by gentlemen who pinned badges to their lapels and called themselves Sheriff or Deputy as the case might be. Arabs, Indians, Pakistanis, Americans, British, even a few continental Europeans, gathered in this multinational assembly to engage in the graft of desert oil exploration, to manufacture a social life that had scant basis in anything save invention, and to sleep, when the night became too hot to stay indoors, under that most awe-inspiring of ceilings, the desert sky.

The drillers were at the centre of this desert existence. They worked with untiring effort and their achievement was immense. They had two rigs at their disposal which had to be dismantled after each well had been completed and then, by an ingenious device invented by the Kuwait oilmen, "skidded" across the desert sands to the next site.

By June 1946, eight productive pre-war wells had been connected to the first gathering centre. They were capable of producing 30,000 barrels of oil a day.

It was a time of rapid development, of ambitious plans and of extreme shortage of supplies. Mr C. A. P. Southwell, KOC's first managing director, joined the company from Anglo-Iranian in 1946 and had the task of reconciling these factors. The need for ingenuity and improvisation was never more apparent than in these early post-war years. Finding sources of supply and keeping them when they were found represented a major international task. The methods which Southwell and his team adopted were not always conventional. In order to enhance the flow of oil from wells to gathering centre, to storage vessels, to sea-loading lines, they bought up an entire disused pipe yard from France. For much-needed workshop accommodation they acquired a hangar belonging to the Queen's flight. Anxious to supplement their medical supplies they purchased a mobile wartime hospital from the Americans, neatly packed in boxes; one box contained hundreds of walking sticks which were resold to the Nigerian Ground Nuts Scheme. To provide temporary power they bought a job lot of ancient electric motors. Most importantly, steel was needed to construct the gathering centre and tanks. These were times of acute metal shortage, and their acquisition of the essential steel was due largely to the co-operation of the Motherwell Bridge & Engineering Company, who not only gave all possible priority to the oil company's requirements but also sent one of its most experienced men to Kuwait as site supervisor to assist with installation work.

Thus, pipelines and a pumping station were installed to convey the crude oil from the wells to the first, improvised gathering centre, where the gas was separated off and the oil pumped to two storage tanks from which it fell by gravity to the coast. To get the liquid from shore to tanker was another matter. A submarine line had to be taken out to water deep enough for tankers to berth.

When a decision as to the siting of a sea loading terminal had finally to be made, officials of the company gathered together in L. D. Scott's office. There was a long debate. The merits of various points on the bay of Kuwait were discussed. The last person to express an opinion was J. A. Jameson. He pointed to a place on the map which stood not in the bay at all but southward between the coastal townships of Fahaheel and Shuaiba. "That is where the terminal will be", he said. And with that he left the meeting. Those who remained said it was unsheltered, it was off centre. They could find little to commend it. But there it was. And looking at the matter with hindsight it is difficult to think that any other choice could have been contemplated. The Dhahar ridge, later to be called Ahmadi, provided a perfect natural descent from the tank farm to that point. More importantly, it was the only part of the coast where the giant tankers soon to be generated by the demands of the oil market and the twists of world politics, could come close enough to be reached economically by submarine pipeline. It was the one segment of coast which would make possible the building of piers large enough to accommodate the massive tankers of the future. JJ, probably the greatest oil engineer of his period, had done his homework; though he may have applied just a touch of intuition to his calculations. It was not long before the greatest fleet of tankers ever to assemble at one time was to line up in testimony to his judgment.

The first twelve-inch submarine line was laid in 1946. It was protected by coal tar enamel, placed on a series of trolleys in line with its ultimate position at sea, sealed at the seaward end to give it buoyancy and then floated out to sea – with the assistance of a tanker.

On 30 June 1946, Shaikh Ahmad al Jabir al Sabah, ceremonially inaugurated Kuwait's oil terminal and sent its first crude oil export on its way. He opened a specially installed silver valve, the tanker *British Fusilier* took on its cargo, and the story of oil in Kuwait began in earnest.

At the ceremony the ruler said:

"Everyone of my people and my friends will rejoice with me in this happy event, which by the grace of God is for our future welfare. I thank God for such an opportunity as this which will help us to continue to carry on various improvements which we desire for the happiness and welfare of my state and people. I wish to mention the assistances rendered to us by the company during their operations

46. Ceremony to mark Kuwait's first oil export, 30 June 1946

47. Shaikh Ahmad turns a silver valve to start oil flowing to the tanker British Fusilier

in my state. Also, I thank His Majesty's Government for their help for the success of the operations, as well as my personal friends the company's directors, both British and American. I also offer my thanks to all the company's staff, all of whom have rendered such valuable assistance. I trust that our friendly relation with the company will continue to exist in a spirit of cordiality and good will."

In his reply, Philip Southwell summed up the occasion and the protracted events that led to it with some delicacy.

"We have met here to commemorate the first shipment of oil from Kuwait, an outstanding event in the development of Kuwait's oil resources which Your Highness entrusted to our company some twelve years ago. . . . The work so far done by the company has had to be carried out in successive stages. In the first place, the company's geologists examined the country in order to see whether, from surface indications, there was any likelihood of finding oil. The second step was to determine whether these surface indications were merely the vestiges of what had once been an oilfield in the remote past, or whether, far down in the earth, there were still large quantities of oil. The company has, after lengthy researches, found that under part of the soil of Kuwait there is a big store of petroleum. The third stage in these operations has now been reached, namely, the actual production of the oil and its transportation to the markets of the world. Your Highness will agree that this successful result of the company's preliminary endeavours presents no mean achievement, particularly as much of the work had to be carried out under the very difficult conditions caused by the war, when the manufacture of essential drilling and other equipment was often greatly delayed and shortage of shipping imposed further limitations on activities. The success that has, so happily, crowned our efforts would never have been possible without Your Highness's unfailing patience, loyal friendship and close collaboration, the excellent work of our Kuwait personnel and the technical skill and large resources of this company, in which British and American interests and personnel are so happily blended."

The Scotts and Donald Campbell had brought with them a magnificent array of "victory" fireworks, supplied by Anglo-Iranian to mark the cessation of the war. And on 30 June 1946 the night sky exploded with rockets, comet trails, shooting stars and the thunder and lightning of a "set piece".

The Dhahar ridge was renamed Ahmadi at this time. It was at a dinner given by the ruler in honour of Anthony Eden, Foreign Secretary in Britain's wartime administration, that somebody suggested they call the new town Ahmadi and the point at which the pipeline came down to the sea *Mina al Ahmadi*, the Port of Ahmadi. The ruler was greatly pleased with the idea. "Anyone who uses the name Dhahar will henceforth be fined 100 rupees," he said.

The five years from 1945 to 1950 represented a period of great change and progress in the affairs of Kuwait. As the oil installations grew, so the working community expanded. As more and more artisans, technicians, administrators and managers arrived, so the associated problems of housing, feeding, education, transport and leisure made inevitable demands on limited resources of goods and manpower. It was not only the oil company that faced the consequence of a non-stop expansion. Kuwait, too, had to implement training courses in the new technology and at the same time import trained personnel to help run schools, hospitals and government services. The country was anxious to put its wealth to the maximum use from the outset; but it soon found that money is of little use without attendant skills.

48. Pioneer oil-driller, Magwa

One of the largest tasks for the oil company at this time was that of obtaining and transporting supplies. By the late forties, it was still a near disaster to lose an anchor in deep water or to break a piece of drilling equipment. The drilling encampments of Magwa and Wara remained more or less isolated, changing only in the growth of their populations and needs, and it was decided to build a residential and administrative town near the rapidly growing storage tank farm six miles from the coast. Meanwhile, Burgan continued to sprout wells in profusion; by 1950, no less than seventy-eight of them were producing at an annual rate of nearly 20 million tons, and twelve rigs were constantly at work adding to the flow. A jetty, the most advanced and massive of its kind at that time, was conceived in 1947 and completed in 1949. A building at Magwa, originally intended as a large-scale living quarter, was turned into the company's first hospital. The office in the Behbehani compound was beginning to burst at the seams. In a Nissen hut nearby a new company medical centre had been set up, limited in size and scope perhaps, but efficient and well thought of by those who had occasion to use it.

A time of transition

Management problems were added to those of supply and political uncertainty in the early post-war period. Local relations, for example, are of paramount importance in Kuwait, whereas they are at most a fringe activity of a vast majority of business firms.

Tom Patrick, who took over operational responsibility from L. D. Scott in 1946, was just as professional and demanding as his predecessor. He became ill in 1948 and went home to America where, tragically, he died from a brain affliction. Briefly, Donald Campbell assumed on-the-spot responsibility as acting general manager and then Aubrey Schofield was sent from London as temporary manager, only to die in an air crash over the desert.

During this crisis of local leadership, Mr Southwell, later Sir Philip Southwell, demonstrated that his interest in the company's affairs was more than that of a head office administrator based in London. He spent a great deal of time in Kuwait. He established an amicable working relationship between the management and leading Kuwait officials. He was in control during a period of massive expansion of output, personnel, amenity and fortune. It was as a tribute to his concern for the Kuwait oil community that the company's hospital in Ahmadi was named after him. In 1948, L. T. Jordan came from Texas to take over as general manager in Kuwait. He was unquestionably the man for the occasion. He fostered a unique community spirit among the oil people of Kuwait, and he enjoyed their total good will. His efforts were complemented by the work of his chief engineer, J. W. Lowdon, the architect of Ahmadi. For ten years from

1946 he was responsible for the major building projects of Ahmadi and of the port and field installations.

Kuwait had yet to confront the affluent fifties. This was a moment of transition. And nothing could have been calculated to make the change more poignantly matter of fact than the death of the country's first ruler of the oil era, Shaikh Ahmad al Jabir al Sabah, on 29 January 1950. Mention has been made of his part in the negotiation of Kuwait's most vital affairs. It is enough to add that he was, during the most difficult period of Kuwait's history, a loyal friend of his Arab neighbours and of Britain.

On 25 February, the Council of the House of Sabah met to elect the new ruler of Kuwait, Shaikh Abdullah al Salim al Sabah. It was a wise choice, for Shaikh Abdullah had played a leading part on the domestic political scene, as head of the Council of State, and virtual comptroller of city and state finances, during a large part of the reign of his cousin Shaikh Ahmad.

49. Shaikh Abdullah al Salim al Sabah, second ruler of Kuwait's oil era

By the time of his succession, the oil company had created a technical base from which production could proceed with increasing volume, efficiency and profitability; it had set up the world's most advanced and effective tanker loading facility; it had made fresh water widely available, created an electricity supply system, built an oil refinery capable of meeting local needs, and set in motion training and health programmes. In the short time in which all this had been happening, the state had devoted a large part of its newly acquired resources to the building of houses, schools and hospitals, and to the formation of long term development plans. It was a good start to an era that held some momentous events in store for Kuwait and the world at large.

Oil from storage tanks at Ahmadi Ridge, conveyed by gravity to the coast, was carried on twelve-inch submarine pipelines – growing in number from the single line of 1946 to ten by 1950 – to a sea loading area a mile from low water mark. This terminal formed the apex of a triangle whose base was represented by the towns of Fahaheel and Shuaiba, and it was from this base line that the south pier of Mina-al-Ahmadi emerged in 1949, with an approach trestle stretching more than 4,000 feet out to sea. With its vast capacity, designed to feed eight of the largest tankers then in existence at any one time, Mina-al-Ahmadi was now able to turn the resources of the Burgan oil field to the fastest possible account. South Pier was, in fact, the outstretched arm of a very significant complex.

The Shatt al-Arab in Iraq and the brackish wells of the desert had been the source of water for the local population for generations and remained so for the first working parties of the pre- and post-war periods. But by 1947 a plan was produced to incorporate a water distillation plant, a power station and a refinery within Mina-al-Ahmadi.

The distillation plant inaugurated in 1949, made up of six evaporator units each with a potential output of 100,000 gallons a day, put an end to Kuwait's dependence on a distant and not entirely reliable source of water. Brackish water from the desert, always in short supply, was subsequently mixed with the distilled liquid to provide necessary mineral salts.

The 22,500 Kw power station of the oil port was another sign of progress. It also signalled the most conspicuous – and certainly the least attractive – of the processes by which industry announced itself in the desert; the spread of a great wirescape, reaching in disarray to the horizon. But visual effects such as these are not confined to Kuwait, and they could hardly be said to outweigh the benefits that electricity has brought to the country.

A KOC refinery was commissioned in 1949, with an initial capacity of 25,000 barrels a day, sufficient it was thought to meet the rising local demand for motor fuel and kerosene, as well as the company's own operational needs. This capacity, based on a limited range of products, was to be multiplied many times over as the demand grew from both company and the state. So too did the extent of the mix which embraced light and heavy gas oils, kerosene, bitumen, gasoline, light distillates, aviation fuels, diesel fuel oil and liquefied petroleum gas.

At this time, Persia was the greatest producer of oil in the Middle East. Then, in 1951, its oil industry was nationalised and the Anglo-Iranian Oil Company lost its major source of supply. A world fuel shortage threatened and Kuwait became, suddenly and acutely, the centre of international attention. In the crisis that followed, Kuwait and the KOC rose to the occasion with a combination of enthusiasm and enterprise which produced some remarkable results, and handsome rewards. In the next two years the production of crude oil was doubled, reaching 37 million tons in 1952. By 1953 the figure was increased to 43 million tons. The statistics for world production at the end of that year showed Kuwait to be the second largest exporter, next to Venezuela, and the fourth largest overall producer. Meanwhile, Persia's industry virtually died on its feet, a single tanker ploughing the oceans in a vain bid to unload its pathetic cargo, the pipelines to Abadan empty and, it seemed, superfluous.

By the end of 1950, a total of 102 wells had been drilled in Burgan and seventy-eight of them were producing while others awaited completion. At the conclusion of another year's intensive work this figure had become 119, with 115 producers; by 1952 there were 126 active wells. At this time, too, eight gathering centres had been established to collect the crude from the wells, separate the liquid and gas components and pass the oil to storage tanks.

Another event of 1951, not unconnected with the Persian debacle, was a new arrangement between the government of Kuwait and the company. The Kuwait Oil Company became an operating concern, managing the installations and working the concession on behalf of the owners; the agreement was extended by seventeen years (expiring in 2026), and new financial terms were negotiated. According to the original agreement, payments from the oil company to the state were based on a royalty of 13 cents-equivalent per barrel. In May 1951 a letter was signed in Beirut modifying the royalty arrangement to a system of remuneration that ensured an equal share of profits, so computed that income tax, royalties, rent and other payments would equal 50 per cent of all export profits. The final draft was signed in December of that year. A clause in this addendum to the original oil agreement stated that the owning companies "renounced all claims that the concession covered the islands of Kubar, Qaru or Umm al Maradim".

In 1948 the Arabian American Oil Company, operating across the border in Saudi Arabia, asked the ruler if it could search for oil on

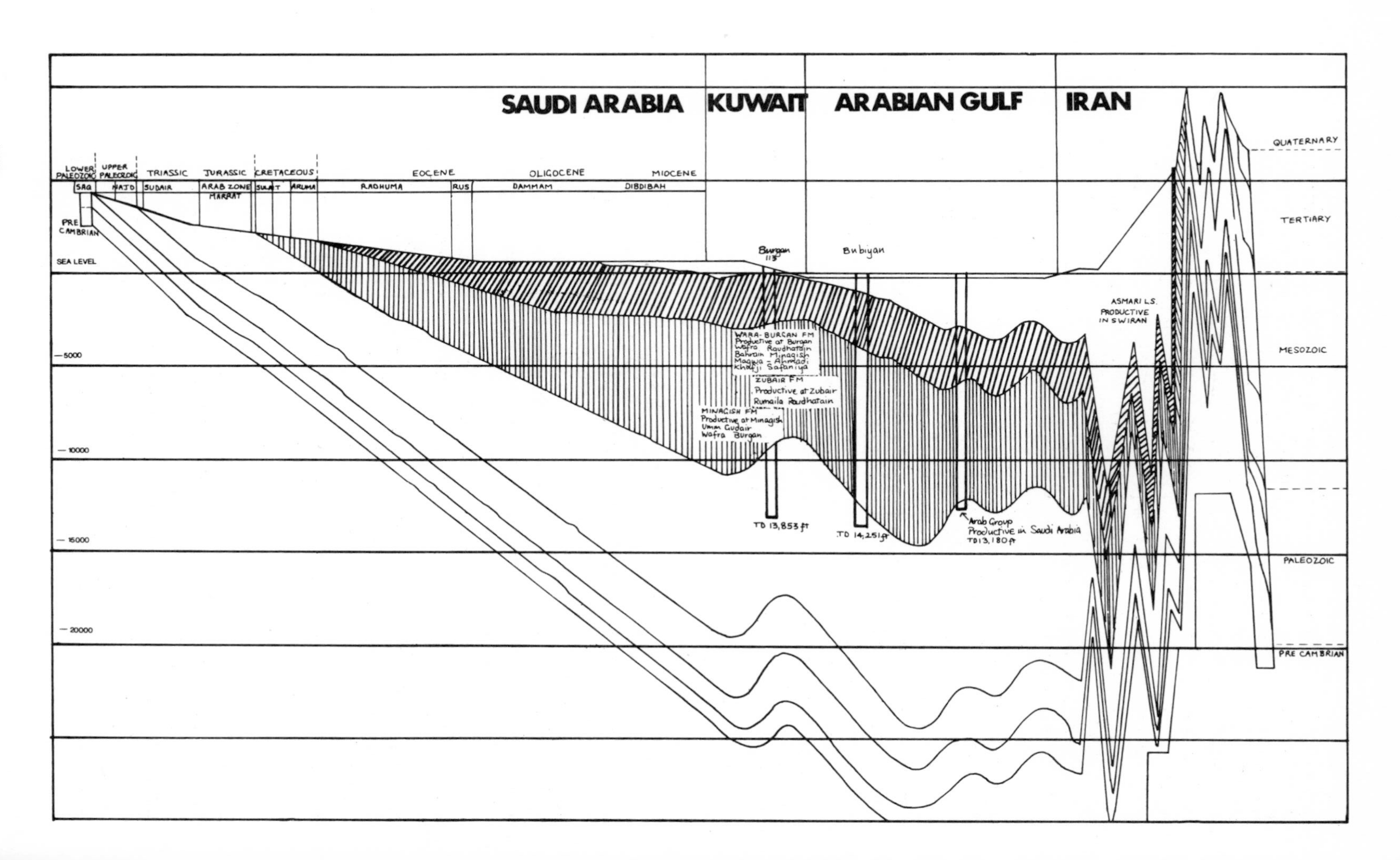

50. *Geological cross section, Arabian Highlands to Zagros mountains, abstracted from KOC drawing*

the islands of Kubar, Qaru and Umm al Maradim. KOC protested that these areas were within the limits of the concession granted it in 1934, and asked that the matter be referred to arbitration. The original agreement had, in fact, laid down that the company should have "the exclusive right to explore, search, drill for, produce and win . . . within the State of Kuwait including all islands and territorial waters appertaining to Kuwait as shown generally on the map annexed hereto". The line of that map extended to approximately three miles from the coast. However, no action was taken pending a definition of the limits of territorial waters and the islands therein. Aramco dropped out of the reckoning and in 1951, KOC relinquished all claims to the islands.

Burgan had become the world's largest single oil field. Structural interpretation of the limestone strata of the area, together with widespread test drillings and further geological surveys, began to suggest that the oil deposits of Kuwait were even greater than had been anticipated. The new agreement reflected the known and prospective wealth likely to accrue to both parties.

In March 1951, Burgan's 113th well was drilled to test the formation below the level at which all the field's production had so far been obtained. The new hole bottomed in the triassic formations at 13,853 feet. In the event, no commercial deposits were found at this level but oil was detected at a higher zone, in the cretaceous oolite, and this formation was subsequently developed. In the same year, a test well was spudded at Magwa and a common oil-water interface with the main Burgan field revealed, thus providing evidence of a considerable extension of the field. At about the same time, the Ahmadi Ridge was drilled in order to investigate the so-called Dammam structural formation. The well began to produce in December 1952 from limestone and sandstone formations. It had been thought that the ridge was of more recent origin than the Burgan and Magwa fields, but from the relative positions of the producing horizons it seems that the Ahmadi Ridge may simply be the result of more erosion in the Greater Burgan than had been suspected. Further drilling at Ahmadi also proved fruitful and helped to delineate the massive extent of the Burgan, Ahmadi, Magwa complex.

In the first full year of the new profit-sharing arrangement, production approached 37 million tons. The effect in monetary terms is discussed in the next chapter. By the end of 1953, a total of 149 producing wells were connected up to eight gathering centres. Of these wells, 130 were in Burgan, but the net was spreading north to Ahmadi and Magwa, and by 1954 the drilling rigs had reached Raudhatain, very nearly at the Iraq border in the north, and Umm Gudair, some thirteen miles west of Burgan.

Along with the activity in the fields which gave rise to this tremendous increase in Kuwait's oil production, went another frantic effort at Mina. Soon after the nationalisation of the Persian fields, the maritime world started to beat a path to Kuwait's door. At one time the armada of vessels waiting to be loaded queued line astern for several miles. Mina and its new jetty were equal to the occasion.

Kuwait's natural assets of geological structure and terrain, and the proximity of its major oil field to tanker loading facilities were an obvious advantage at the time of the Abadan crisis. At the control room in Ahmadi, electronically operated valves were set to govern the flow of oil from the gathering centres of Burgan to the tank farm and on, via the gravity flow lines, to the South Pier and the ever-growing formation of tankers waiting to tie-up at its berths. At the same time, the pump station control room directed a constant flow of bunkering fuels to this international assembly of vessels. The unloading of general cargo was an administrative and physical task sufficient to tax the resources of any terminal in the world. It was, of course, an emergency operation, far removed from the formal procedures of advance warning and nomination of cargoes which apply normally.

A world at first barely interested awoke to the fact that Kuwait was capable of producing, from one partially developed field, with 120 operational wells, a remarkable volume of oil. Something like 40 per cent of the country's annual production was exported to the United Kingdom, 17 per cent to France, 16 per cent to North and South America, the rest to a growing number of markets throughout the world. The demands created by the closure of Abadan had an undeniable influence on the development of Kuwait as an oil-producing country. The government had every reason to encourage a massive effort to meet these new demands. An unexpected force had accelerated the growth of oil production and given added zest to other, associated programmes. Kuwait stood at the verge of a golden era. Planning had become the keyword.

The first wildcat was spudded at Raudhatain in September 1954 and reached a depth of 10,301 feet. It was completed in the Zubair sand formation and substantial oil deposits were encountered at other levels – in limestone strata of the Ahmadi shale and in the Mauddud formation. In 1956, three further wildcats produced evidence of one new field and two productive wells. These were at Bahrah, scene of the first drilling operation in Kuwait, where oil came up from the Burgan sand formation at 8,488 feet; Sabiriyah, completed in the Mauddud formation, although additional accumulations were found in the Burgan and Zubair formations; and Magwa, where well No. 41 showed oil in the Wara sand which, though not commercial in itself, led to profitable drilling later on. Another test well, at Mutriba, was spudded in mid 1957 but was abandoned after reaching 11,516 feet without a show.

The strange reasoning which led Sir Percy Cox to hive off from Kuwait territory the area which came to be known as the Neutral Zone, was shown to be prescient – though whether it was justified is another matter – by the discovery of oil in the Wafra field in March 1953. Kuwait and Saudi Arabia shared equal rights in this zone. In 1947, Shaikh Ahmad had said that his country's portion would go to the highest bidder.

On 28 June 1948, Kuwait's concession was in fact granted to the American Independent Oil Company (Aminoil), a consortium formed on the initiative of Ralph K. Davies, a former vice president of Standard Oil of California. On 20 February 1949, the Saudi Arabian rights were assigned to the Pacific Western Oil Corporation which later became the Getty Oil Company. This time the currency of negotiation was the dollar. The Aminoil contract involved a down payment to Kuwait of 7·5 million dollars and an annual payment thereafter of 625,000 dollars. Royalty was fixed at 2·5 dollars a ton, while Kuwait reserved the right to one-eighth of local profits made from production and refining.

The Pacific Western contract with Saudi Arabia for its undivided half share involved an initial payment of one million dollars, and a royalty of 55 cents a barrel. The government was to be paid one-eighth of profits from production for local use and a quarter of the profits derived from refined products. Like the east Arabian concessions, Major Holmes and the Eastern and General Syndicate had allowed its Neutral Zone option to lapse. It was unfortunate for the untiring Holmes. It was not for want of trying on his part, or on the part of his colleague T. E. Ward, that the prolific oil fields of the area went begging for so long. But he had the satisfaction of knowing in the end that Kuwait, in whose concession he retained a stake, was the most prolific of all.

Aminoil was made the operating company for the Neutral Zone. Between 1949 and 1953, it drilled five dry holes before striking oil at Wafra. The first full year saw a production of 2·8 million barrels. After five years it reached 14·7 million barrels, which compared with 509·4 million barrels from the KOC fields. It is a relatively modest field, but not to be sneezed at for all that. In 1958, Getty completed a refinery with a capacity of 50,000 barrels daily at Mina Saud in the zone, while Aminoil built a 30,000 barrel capacity refinery at Mina Abdullah in Kuwait.

The name of the British signatory to the Kuwait concession was changed again in 1954. The Anglo-Iranian Oil Company became British Petroleum. That company had fought a long battle in the courts with Britain's tax authorities, the technicalities of which bore on the need for the revised Kuwait Oil agreement in 1951. In 1955, a decision was made to use the "posted" price of oil, that is its estimated value at the post-production stage rather than the price it is actually sold at in fluctuating market conditions, as a basis for calculating revenue. The royalty figure was fixed at 12·5 per cent of the posted price. Thus, an amendment was made to the 1951 revision and a bonus of £5 million paid to the Kuwait government.

In 1956 came another of the gales which every now and again blow through the Gulf, though they often lived up to their reputation of doing somebody a good turn. This time it took the form of the Suez escapade. Whatever the military and political arguments for operations such as this, the economic effects are unquestionable. Kuwait, like other oil-producing countries in the Middle East, is highly sensitive even to the threat of change. When its supply routes are endangered price increases can be incurred which result in fluctuations in the revenue of both Kuwait and the owning companies, to say nothing of the major consumer countries. Other oil-producing countries suffered from the same shock wave. The Organisation of Petroleum Exporting Countries (OPEC) was formed in 1960 in an effort to deal with the effects of outside influences on oil prices and thus on the economies of producing countries. Kuwait joined this organisation at the time of its inception at a conference held in Baghdad from 10 to 14 September 1960. Kuwait, Iraq, Iran, Saudi Arabia and Venezuela were the founder members. Kuwait's representative held the post of Secretary General of OPEC from 1 May 1965 until the end of 1966. On 9 January 1968 the Organisation of Arab Petroleum Exporting Countries (OAPEC) was set up by an agreement signed in Beirut by Kuwait, Saudi Arabia and Libya. Its headquarters are in Kuwait and from the time of its establishment, OAPEC has had a Kuwaiti official as Assistant Secretary General. These bodies have helped to prevent unnecessarily drastic price fluctuations, and they have certainly given to the producing countries a powerfully concerted bargaining power, which those countries have shown little reticence in using.

The Suez intervention may have had all kinds of repercussions, but it did nothing to stunt the growth of the oil industry in Kuwait, although production temporarily levelled off as a consequence of tankers having to make the long haul by way of the Cape.

By 1957, KOC's production had risen to over 56 million tons, an increase of more than two million tons on the previous year, although Suez was not reopened to tankers until mid-April. Eleven gathering centres connected 238 wells, with a combined throughput of 1,139,850 barrels a day. Two new crude oil distillation units were added to the refinery. Each had a capacity of 80,000 barrels a day, thus making possible by 1958 a total daily output of 190,000 barrels to provide for a rapdily increasing local demand for petroleum and other distillates. It processed almost exactly 10 million barrels (1,348,698 tons) during the year, as well as 13,806 tons of bitumen. The local demand for motor fuel rose in the same year, 1957, by 37 per cent over the previous twelve months, reaching 24 million gallons.

By this time, 53 per cent of all oil industry employees were Arab, a growing number of them Kuwaitis. The KOC Training Centre in Magwa, opened in 1951, had provided a wide range of courses for a total of 1,773 Kuwaiti students by the end of 1957, and of this number 954 remained on the KOC payroll. Some were on the way to supervisory status.

Marine exploration rights off the Neutral Zone, outside territorial waters, went to the Japanese-owned Arabian Oil Company in 1958, by agreement between Kuwait and Saudi Arabia. This concession involved separate contracts between the Arabian Oil Company and the Shaikh of Kuwait and the King of Saudi Arabia, for their respective half interests. Each agreement called for an annual rental of 1·5 million US dollars, plus one million dollars on discovery of oil, to be backdated to the inception of the contract. Each also provided for a minimum annual royalty of 2·5 million dollars. The agreements differ, however, in that Kuwait was to receive 57 per cent of profits calculated on posted prices, and for any other profitable activities, while Saudi Arabia would receive 56 per cent of net income. Both countries took up a 10 per cent shareholding in the oil company.

The Japanese company's first drilling effort was disastrous. Gas was struck unexpectedly at 1,507 feet in August 1959, and the well burned fiercely for nearly two weeks before it was quelled by American firefighters. Drilling was resumed after three months, and in January 1960 the second test well began to flow from 4,900 feet. The new submarine field became known as Khafji. By May 1961, fourteen wells were producing from the cretaceous structure, at the rate of about 6,000 barrels a day. Permanent storage and loading facilities at Ras al Khafji in the south of the Neutral Zone, with a capacity of 2 million barrels, were completed in 1962. By 1969 annual production was to reach 16,534,071 tons, nearly three times the onshore Neutral Zone figure.

On 10 February 1959, the Chief of Public Security, Shaikh Jabir al Ahmad, was made President of Kuwait's Finance Department. With that appointment Shaikh Jabir became the overlord of Kuwait's oil affairs, giving a new impetus to the government side of oil relationship and preparing the ground for a highly efficient administration in the very eventful years that were to follow. In 1962, with the new Kuwait constitution, the department became the Ministry of Finance and Oil.

In little more than a decade, production had grown from under three million to over 80 million tons a year. In 1950 there were seventy-eight productive oil wells; by 1960 nearly 400 were in operation. New fields had become linked to the distribution grids; new names, or new at any rate to the outsider and the compiler of oil records – Minagish, Raudhatain, Umm Gudair, Sabiriyah, Magwa, Ahmadi, Wafra and Bahrah – were on the map.

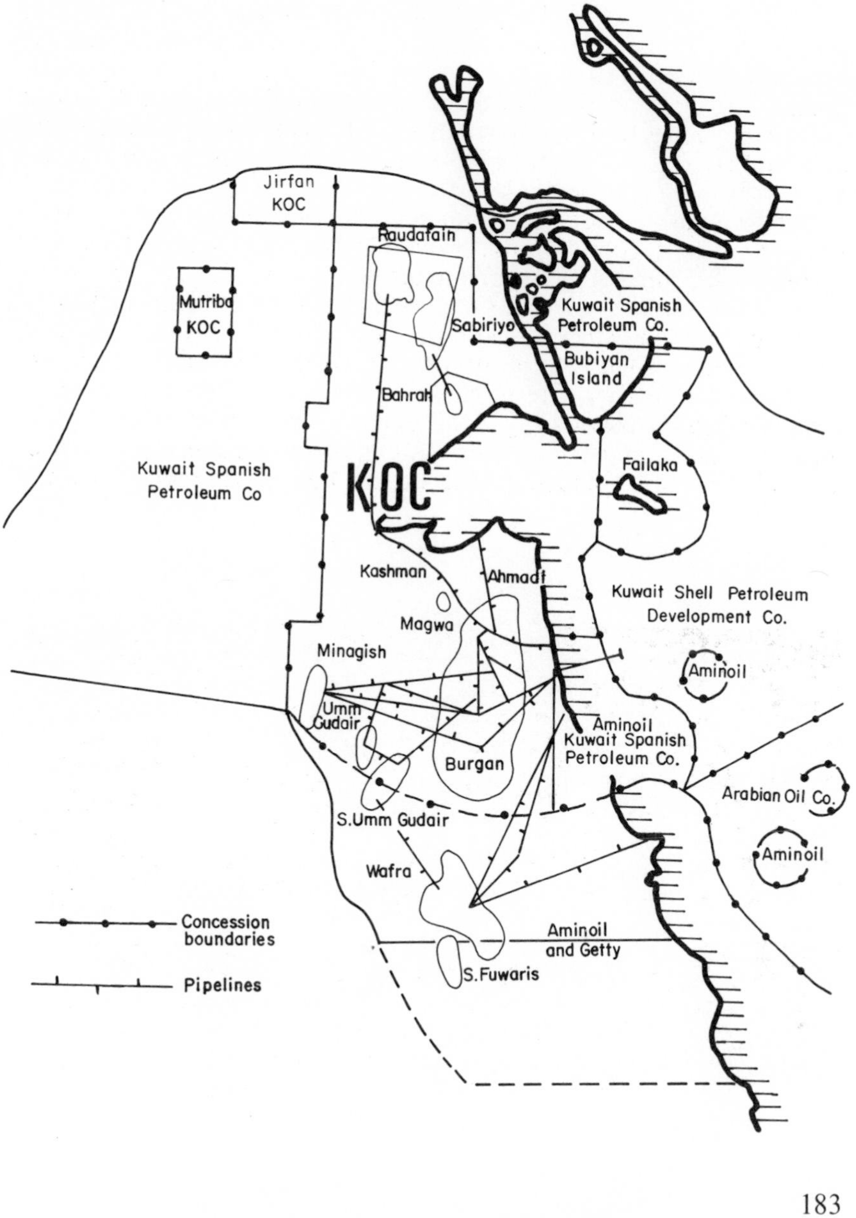

viii Kuwait's oilfields, after Buchanan

Oil community

Along with the emergence of Mina-al-Ahmadi and its ramifications, a dramatic change was occurring on the ridge above. A gradual evacuation of the spreadeagled dwellings of the Magwa and Wara sites had begun in 1947, gaining momentum until, in the early fifties, the town of Ahmadi was established. Unlike Magwa and Wara, Ahmadi brought a pleasantly ordered suburban togetherness to this desert site. In conception it was profoundly English, though its numerical streets and avenues and its Scandinavian and American architectural touches combined to make it acceptable to all the nationalities who worked thereabouts. By 1952 there were some 350 staff houses and 700 payroll employees' houses, built at a cost of about £$3\frac{1}{2}$ million, arranged in precisely delineated streets, cut through the centre from east to west by 7th Avenue. Thus everyone – Arab, American, European, British – could feel at home. And make themselves at home they did. The first foreign tree (flown from Basra) was planted in March 1948 and subsequently sandy stretches which for hundreds, indeed thousands, of years had supported nothing more demanding or colourful than the desert's arfaj plant, began to sprout evergreen hedges and tree-lined streets. It was not long before white and timbered house fronts were hedged with greenery and framed by a profusion of flowers.

But Ahmadi was not just a dormitory suburb. It was, and remains, the hub of the oil community, providing for health and welfare; for all the social, cultural and sporting facilities of an advanced and well-off, multi-national work force. It added a great many of the tasks of a municipal authority to KOC's existing responsibilities. Road building and maintenance with all the attendant problems of traffic flow, intersection design, and long-term planning, became an increasingly large task. By 1955, a swimming pool, a local ice-making plant, public security headquarters complete with their own mosque, a post office, a laundry, customs offices, a bakery and a bus service represented the

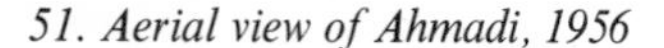
51. Aerial view of Ahmadi, 1956

most conspicuous parts of an impressively mixed bag of public utilities and social amenities. In 1947, a Swedish style house served as an Anglo-American school for six pupils. By 1961, there were 401 pupils on its register.

The constructional tasks involved brought about an acceleration of the policy of employing local Kuwaiti contractors who, by 1955, carried out more than 90 per cent of the work on these expanding civic and industrial programmes.

Clubs, catering for the social and sporting requirements of the community, and for their differing languages and ethnic needs, sprang up in Ahmadi, with names like Hubara, Unity, Nadi al Ahmadi and Nakhlistan. They formed a yacht club, soccer, rugby, cricket, hockey, sailing, darts, bridge, and, of course, gardening clubs. They competed with each other and they began to go abroad to Saudi Arabia, Lebanon and Cyprus to do battle. Kuwaitis showed a ready aptitude for the game of football, and some prowess at snooker, though it would have been to the greater delight of many Englishmen on the spot if one of them had displayed a respectable cover drive. A community hall provided a centre for lectures and meetings and a home for a versatile and enterprising theatrical company.

When it reached the peak of its development this remarkable township – planted with all its twentieth-century attributes on an arid strip of desert – comprised some 3,000 private residences, a three-block guest house for company visitors, ten blocks of bachelor flats, and a shopping centre complete with supermarket. It also contained seventy-six miles of made-up road, three sewage pump houses, a water storage capacity of 18,700,000 gallons, a central workshop and countless refrigeration, air-conditioning, water-cooling and associated units.

In 1952 Ahmadi and its associated installations consumed 2,510 million cu. ft of gas, routed from the oil fields. By 1970, the figure for KOC consumption alone was 79,000 million cu. ft. In electrical power

52. Aerial view of Ahmadi, 1968

generation, too, the same kind of growth typified the magnitude of the engineers' contribution to the development of the oil community and to the technical advancement of the state.

53. Trees in the street, Ahmadi 1970

Further field development

The flux in oil and international affairs which characterised the fifties carried over into the next decade.

But Kuwait is cushioned against extremes of economic change by the near miracle of its geological history. The abundance and accessibility of its oil enable production costs to be kept at a remarkably low level. For twenty years or more after the country's first export cargo, Kuwait's production costs remained at about one third of the Middle East average, and one twentieth of the US figure. But these advantages must be equated with the need to transport the oil over vast distances by tanker. Even with the canal open, the Gulf states are not especially well placed in relation to the major European and American markets. With Suez closed, transport costs inevitably go up with consequent price rises. And in the modern world, price rises, once made are seldom reversed.

Kuwait is also subject to international alliances the aims of which are not always consistent with the provident exploitation of reserves or the maintenance of a competitive position in the market place.

Such considerations apart, Kuwait entered with confidence a period which was to witness a world-wide explosion of productivity and wealth. By 1960, the whole of the state had been covered by reflection traverses at ten kilometre intervals. The oil company was thus able to map out a long-term programme for further exploration and for exploiting existing productive zones.

In 1961, another offshore area came into prominence. This was on the shelf of the Gulf off Kuwait's coastline, some six miles from low tide mark. Sited directly between the very rich fields of Kuwait itself and Khafji, the exploration rights were keenly sought. They went to the Kuwait Shell Petroleum Company, and the terms of the agreement closely followed those of the revised Kuwait–KOC document. Additionally, the Kuwait government reserved the right to purchase an interest of up to 20 per cent in the company and to appoint two directors to its board. In the event, this was one of the few Gulf exploration programmes to draw a blank. After nearly two years of drilling the Kuwait Shell Company gave up its quest in October 1963. It was said to be awaiting clarification of maritime boundary lines.

In 1962, KOC voluntarily gave up about half the area covered by the original concession agreement, including the whole of the western portion except for the Mutriba and Jirfan areas. These relinquished regions were leased to the state owned Kuwait National Petroleum Company who subsequently formed a partnership with Hispanoil.

Exploration of the fields retained by KOC went on with undiminished success. The Minagish field, where the discovery wildcat well was spudded in October 1958, had shown substantial deposits at various levels – in the Mishrif, Wara and Burgan formations of the middle cretaceous, in lower cretaceous oolite limestone (called Minagish oolite in Kuwait). By 1963, other wells in this region showed that there was oil saturation at the jurassic level of the Arab zone from which Saudi Arabia's oil derives. By 1967, twenty-three wells had been drilled in Minagish, four of which were converted for gas injection to maintain the pressure in the oolite reservoir. The discovery well was drilled to a total depth of 11,614 ft in the Hith formation but was plugged back to the pay zone at about 10,000 ft.

At Dibdibba drilling began in 1959 and continued until 1962, but all the wells, which bottomed in the Burgan sand formation at 8,500 to 9,438 ft, were dry.

In 1960, exploration of the Mutriba field was resumed. An earlier drilling between 1957 and 1959 had failed to show oil in commercial quantity. No. 2 well was no more encouraging. After penetrating through to the lower depth of the Arab zone at 15,600 ft, without a show, the well was plugged in April 1962. A third well was spudded in August 1965, but this too was dry. Some very low gravity crude was found at the Burgan level, but in neither quality nor quantity did it justify further exploration. Similar blanks were drawn at this period in the Mityaha area. But these were small disappointments in a programme of discovery and development which throughout Kuwait's oil history was marked by an embarrassment of riches.

In 1961, further investigation was made of the already productive Raudhatain field. Tests were carried out to a depth of 12,983 ft at the Minagish oolite level and at intermediate depths in the Ratawi, Shuaiba, Zubair intervals. Well No. 38 was completed in the Ratawi limestone at 11,360 ft but was later recompleted in the Ratawi sand. Raudhatain 42 suggested that oil was present at favourable pressure at lower levels. Meanwhile, further productive wells were drilled at the Burgan level in this northern field. By 1969, fifty wells had been drilled of which only three were not completed. Further hydrocarbon accumulations have been identified at lower levels than those which have so far been exploited and wells have been completed in all the major reservoirs, though some minor reservoirs could yet prove worthy of development.

Between 1961 and 1967, an extensive programme of seismic and drilling investigations was carried out on the periphery of the Burgan field. These confirmed the existence of heavy oil and bitumen deposits in the Dammam, Rus, Radhuma and Tayarat formations, but in most cases it would be uneconomic to attempt to recover it. Other test and wildcat wells were drilled to examine the oil bearing qualities of the Minagish oolite, Ratawi limestone and Zubair sand formations. Although oil-bearing zones were penetrated, especially by Burgan No. 343 which was completed in the oolite limestone, most of these wells were plugged back and completed in the Burgan sand zones.

Investigation of the Umm Gudair field began in 1962. This followed the discovery of the Minagish field in 1958, and the interest raised by its oolite formation which, it was now evident, extended towards Umm Gudair and Burgan. A full appraisal was carried out between November 1966 and January 1968, with the drilling of twenty-seven wells. In addition, the Aminoil Company drilled seventeen wells in the southern extension of this field. The reserves of the area vastly exceeded the expectations of both companies. By the end of the programme, KOC had drilled thirty-two wells in the separate east and west structures of Umm Gudair. The western structure seems to connect with the Wafra field in the Kuwait Neutral Zone, while the eastern oil-bearing formations appear to connect with the southwest flank of the Burgan field.

At this time, Bahrah was also looked at more closely following the successful completion of well No. 2 in 1956. Seven new test wells were drilled, two of which penetrated the Ratawi limestone formation. The others were intended to evaluate the Mauddud limestone and Burgan sandstone formations. The productive intervals were of a thin and

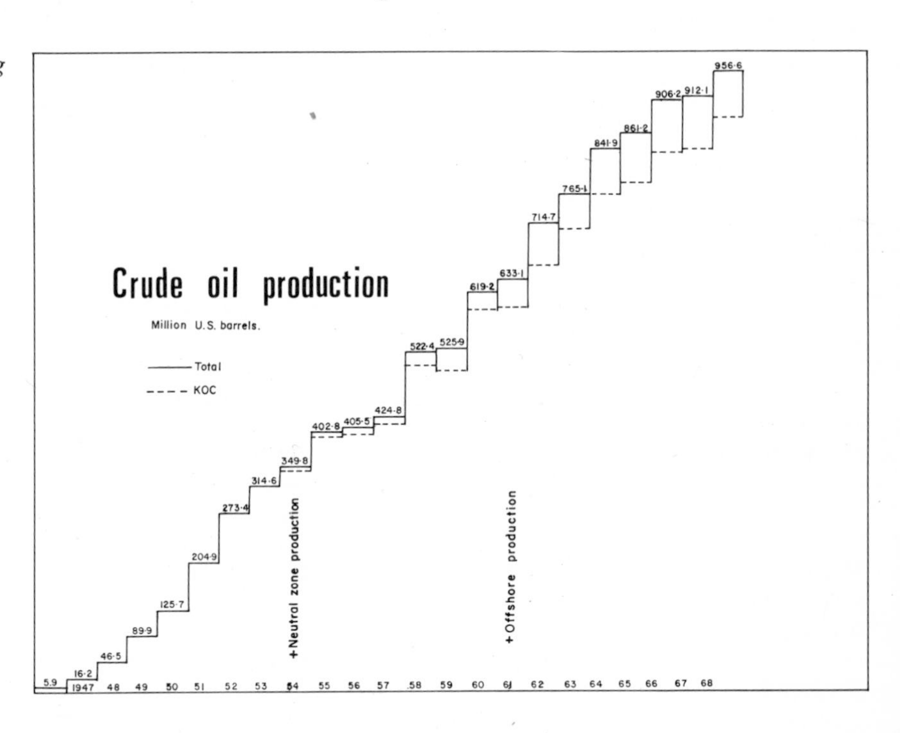

54. Crude oil production showing KOC and total Kuwait figures

erratic nature which made thorough evaluation difficult. It is thought, however, that early geological prediction of oil in Bahrah may well be vindicated by further investigation. There are indications that commercial quantities may be present in the Mauddud and Burgan formations, though they will probably require an artificial lift. There are also deep structures of considerable interest at the permian and carboniferous levels.

Other exploratory work carried out in the early 1960s, included directional drilling from an off-shore platform off Medina in Kuwait Bay. Work began on 9 September 1962, with Medina No. 1, which flowed at the rate of 641 barrels a day from the Burgan level. Medina No. 2, suspended in September 1963, produced at the rate of 1,870 barrels a day from the Wara level. The third well produced a small amount of hydrocarbon from the Ahmadi formation but it was of no economic significance. If the Medina structure, much of which is under Kuwait City, were to be developed it would be necessary to drill obliquely from a remote location. It is unlikely, therefore, to be exploited for some time to come.

Several wells were drilled at Sabiriyah from 1963 to 1965, some being completed in the Burgan and some in the Mauddud formations. They were intended to define the various structures of the field and to delineate its limits. All known producing formations in Kuwait were encountered in this field. By the end of the sixties, twelve wells were completed from the Burgan formation, thirteen from the Mauddud and one from the Ratawi sand. In addition, several development wells were drilled and subsequently plugged. Considerable reserves are known to exist in the Zubair zone but so far no producing wells have been completed in this formation. There is also a considerable saturation of the Tuba limestone but the oil is difficult to recover. Electric log tests have shown a close correlation between the Sabiriyah and Raudhatain fields, with only minor variations in formation thickness and lithological make up.

Other development work carried out in the period 1960–5 must be summarised in order to bring up to date the picture of the intensive and uniquely successful exploitation of Kuwait's oilfields.

In 1962 a wildcat at Kashman No. 1 pointed to the possibility of a new field north-west of Magwa. After testing at the Ratawi limestone level, the well was completed in the Wara sand, producing 1,022 barrels a day. Further test wells showed this to be a disappointingly small reservoir, however, and it was not proceeded with. The oil from the upper Wara zone is relatively sulphur free, and this field could justify further exploration at a later date.

The Magwa-Ahmadi region was further examined from 1963 to 1965. A test well at Magwa was intended to investigate the Minagish oolite sequence in relation to the Burgan field ten miles away. In fact, there was no equivalent Minagish development and the well was completed in the Burgan formation. Another well in this region, Dhahar No. 1, though non-productive, provided structural evidence which resulted in a considerable enlargement of the Magwa-Ahmadi fields. Another attempt was made to investigate production possibilities within the city of Kuwait, but the only likely formations, the Wara and Burgan, were found to be water bearing. Bubiyan Island, the barren pancake to the north of Kuwait Bay, and Failaka Island, both proved disappointing.

There is a tenet of oil husbandry which says, arbitrarily perhaps, that production in any one year should not exceed one-twentieth of reserve. The exact reserves of Kuwait are not known. They are said to be the largest in the world and have been estimated at over 10,400 thousand million tons, or nearly 17 per cent of all known reserves; enough to continue the present production rate for almost a hundred years. Thus, the oil companies are working well within a margin of prudence. By the end of 1963, a total of 493 wells had been spudded, of which 445 were connected as producers, and only thirty-two had been abandoned altogether. Can any other oil producing area have shown such fertility?

The procession of broken records – bigger, wider, deeper, higher accomplishments, in oil production, in refining, in the development of Kuwait itself – went on through the sixties. No analogies or figurative descriptions, it seemed, could exaggerate the mounting achievements of the place – and few were spared by those who wrote and talked about it over the years. In 1964, cumulative crude oil production and export exceeded 100 million tons. It was appropriate that the 100 millionth ton (excluding Neutral Zone, offshore, and KNPC-Hispanoil production) was loaded by Kuwait's own giant tanker *Ahmadi*. KOC refinery capacity was increased in 1963 from a throughput of 190,000 to 250,000 barrels a day, and subsequently by stages to 280,000.

The commissioning of another jetty in 1959, the North Pier, had raised the loading capacity of Mina-al-Ahmadi to more than two million barrels a day. The construction of a massive thirty-inch pipeline to link the northern gathering centres of Raudhatain and Sabiriyah with the tank farms of Ahmadi, meant an even greater flow of oil from well to tanker, and at Mina-al-Ahmadi the average turn around time reduced from 40 hours 49 minutes to 24 hours 49 minutes.

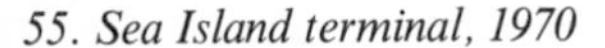

55. Sea Island terminal, 1970

Kuwait was ready to take a bigger share in the management of the country's major source of wealth. In 1960, the ruler appointed two Kuwaiti citizens to the Board of KOC, Sayid Mahmoud Khalid al Adasani and Sayid Faisal Mansour Mazidi. The year before, Sir Philip Southwell retired from the managing directorship. Mr J. M. Cooper, who had played an outstanding part in the affairs of the company as deputy managing director since 1948, also left the board. In 1959, the Hon. W. Fraser (the second Lord Strathalmond) became managing director. L. T. Jordan was appointed deputy managing director, and in the same year the contribution that this remarkable Texan made to the advancement of Kuwait's oil industry and to the development of Anglo-American understanding received acknowledgement. He was knighted in October 1961.

Ambitious plans

At the end of 1964, on 24 November, the second amir of the oil era, Shaikh Abdullah al Salim al Sabah, died at the Dasman Palace in Kuwait. His predecessor, Shaikh Ahmad, was in reality the last leader of the old Kuwait. It fell to Shaikh Abdullah al Salim to implement a new order in which the concept of a modern welfare state was pre-eminent. He embraced the ideas of universal suffrage and social progress with enthusiasm. Few men who have looked on the prospect of unlimited wealth and unrestrained power could claim as much.

Shaikh Sabah al Salim al Sabah, twelfth ruler of Kuwait from the accession of his namesake in 1756, third ruler of the oil age, became the first head of state to be elected under the new democratic constitution. His experience of government administration, of foreign affairs and of the political complexities of the outside world, were more than useful assets as Kuwait went about the consolidation of its past gains and began to take its place on the international stage. Such qualities were reinforced by a government of wide-ranging abilities, led by the heir apparent, the able and very decisive Shaikh Jabir al Ahmad al Jabir.

The most remarkable of the plans put in hand at this time was one for a deep-water ocean terminal, a man-made island, to cope with tankers which, like unconstrained giants, became more bloated each year. The surprises which lay beneath Kuwait's surface were by no means all revealed, either. Continuing seismic and drilling surveys showed yet further extensions to the oil zones. Fail-safe devices were introduced at gathering centres, enabling an automatic shut-down in an emergency. They added to the already considerable safety precautions built into the oil fields. Kuwait's record in this respect had been outstanding, though two fires did occur. The first, in 1964, demonstrated the hazards that can confront the drilling teams. While they were spudding well No. 331 in Burgan field there was a blow-out which sent flames and acrid smoke hundreds of feet in the air and produced a heat of terrifying intensity. It illuminated the desert sky with a ferocity that few witnesses will forget, burning for forty days until, finally, it was extinguished. There were no casualties, but it was a salutary experience, even though another conflagration was to break out in 1971. However, this second fire took less time to extinguish.

Natural gas, as preposterously abundant as oil in the rich underground reservoirs of Kuwait, was used by KOC to power turbine generators at Burgan from 1946, thus converting part of this vast reserve of energy into electricity for company and state use. But the quantities of gas involved could not be even marginally used up in this way. In 1957, for example, production of gas amounted to the astronomical figure of 195,000 million cu. ft. It was possible to use 10,000 million cu. ft or so for power generation. A scheme was set afoot in 1959 to use compressor plant at the Burgan Power Station to inject surplus gas back into the oil producing sand formations of the field. The £1½ million plant, commissioned in July 1961, compressed and injected up to 100 million cu. ft of natural gas a day. But many wells have operated in Burgan for more than twenty years, giving up their gas but losing no appreciable pressure. In 1965, a similar plant – capable of injecting up to 50 million cu. ft a day – was commissioned at Raudhatain in North Kuwait, and in July 1967 a 150 million cu. ft/day plant went into operation at Minagish.

Another and more immediately profitable use for gas is to compress and fractionate it, removing the propane and butane layers as liquefied petroleum gas, or LPG. Construction of a plant to take gas from three of the Burgan gathering centres was begun in 1960 and completed in 1961 at a cost of £3 million. Regular bulk exports of propane and butane began in 1962. At the refinery also a plant was installed in 1961 for the treatment and storage of LPG. With the emergence of large refrigerated tankers, LPG became a valuable addition to the country's exports, Japan taking the major share of its output.

By 1968, the gigantic and imaginative Sea Island terminal, commissioned by Shaikh Sabah the following year, was serving the new generation of 326,000-ton supertankers. Crude was fed to this man-made island, ten miles out at sea, by the largest diameter submarine oil pipeline ever laid. Oil from the northern KOC fields was carried along a seventy-seven-mile route to the northern tank farm by a 30/36-inch pipeline. More storage tanks, more gathering centres had been completed. In 1970 crude output from all fields reached 137·5 million tons. By 1971 production by KOC alone amounted to 144·5 million tons, or 1,067·8 million barrels. And Kuwait together with its OPEC partners, began to look for more revenue from its primary asset.

Area	Producible wells	Wells drilled
Burgan	368	386
Magwa	100	107
Ahmadi	78	82
Raudhatain	51	58
Minagish	14	23
Sabiriyah	48	51
Umm Gudair	31	32
Bahrah	2	7
Total	692	746

Status of KOC wells in producing fields 1970

56. Aerial view of KOC refinery and storage area with Shuaiba industrial complex in the background

8. Politics and Economics

Comparison between the economics of pre- and post-oil Kuwait is meaningless. The two periods are of totally different dimensions. Nevertheless, it is interesting to look briefly at the commercial scene as it was in the days when pearls and animal skins represented the country's primary direct sources of income; when, in demonstration of Kuwait's instinctive sense of commerce, the harbour bristled with dhows whose capacities were measured in terms of the date packages they could accommodate, and which plied the oceans filled mainly with cargoes of re-exported products such as sugar and spices.

Colonel J. C. More, in his 1926 trade return to the Indian government, provided a statistical summary.

The capital, he said, had a population of some 50,000 souls. About 10,000 of these were Persians, 4,000 Negroes, a few Jews, and two or three Chaldean Christians from Iraq. The rest were Arabs. Outside the city only Arabs, mostly nomads, were to be found. The population of Jahra he estimated to be 500. The exchange rate at that time was Rs15 to one pound sterling. A new road had been constructed between Kuwait and Zubair and the 150-mile journey along its entire length cost the equivalent of 55p. by car (with seat), 49p. by van. Imports, mostly from India, Iraq and the Gulf coast, were shown for three annual periods: 1923–4 (£625,064), 1924–5 (£495,972), 1925–6 (£448,014). Rice and coffee were by far the largest items. Exports, mostly delivered to Persian ports, for the same periods amounted to: 1923–4 (£276,541), 1924–5 (£546,651), 1925–6 (£246,746). Hides, pearls, re-exported sugar and tea, were predominant. The figures are suspect especially with regard to income from pearling but they provide useful trade comparisons.

Kuwait in those days was not a rich country, but neither by Middle East standards was it poor. It made the best of meagre natural resources, supported itself through two world wars and the critical period of the Saudi blockade, and boasted a number of wealthy merchant families. Yet it remained an essentially tribal society, without book-keeping or a civil service to record and regulate its finances, its women in purdah, its feet planted firmly in desert tradition. In fact it was not until the mid 1950s, some seven or eight years after the first oil export, that the economy generated perceptible changes in social and administrative structure. There had been a number of earlier attempts to introduce modern amenities such as the setting up of a health ministry by Shaikh Ahmad in 1936 and the opening of a free clinic to supplement the strained resources of Dr Mylrea's Mission. There was even a plan to build a new hospital, but it foundered for lack of money and materials. New building, the setting up of government offices such as the Kuwait Development Board, schemes for expanding industry and the social services, began in earnest by about 1952, gradually gathering momentum until, by 1960, they had brought about a remarkable change in the appearance and the political and social make-up of the country.

The concept of the welfare state is part and parcel of the paternalist nature of Kuwait's shaikhdom, though it owes a great deal in inspiration and form to Western models. As soon as money became available in any volume, health and educational programmes had a prior call on it. The Emiri hospital, completed in 1949, was the first step in a social services programme which has no parallel of scale anywhere in the world. Soon, mental hospitals, sanatoria, maternity units and a new general hospital were under construction. The most advanced diagnostic and therapeutic equipment available was shipped to Kuwait in enormous quantities. One of the first objectives of this expanding health service was to deal with the two most serious endemic diseases of the desert, smallpox and tuberculosis. The American Mission hospital had fought a long and difficult battle to keep the former under control. But its efforts were hampered by the secrecy to which families were driven by fear of the disease. Smallpox was mainly brought in by tribesmen from the Najd and the last great epidemic had occurred in 1932 when refugees crowded into Kuwait following the Ikhwan rebellion. There were sporadic outbreaks after that, but few victims ever found their way to Dr Mylrea's clinic. They were simply left to die or to recover, and many older Kuwaitis today bear the scars of fortuitous cure. Of the 1932 outbreak and its consequences, Violet Dickson has provided a graphic description:

"In the first ten days of the epidemic over 4,000 persons died. It was a terrifying sight to see the corpses being carried daily to their last resting place. But so great was the secrecy that we could not find out if cases had occurred in the houses of our own servants. . . . Among (those) who fell victim was Nazaal, the ruler's chief guide and falconer, and great was Shaikh Ahmad's distress. The primitive Arabs knew only one supposed cure; they believed that one particular smell, different in every case, had the power to cure, but the problem was to find the one and only smell which would be effective. In Nazaal's case every possible thing was brought before him, fruit, flowers, vegetables, cooked food, etc. Then children, young women, and old women were made to pass before him. . . ."

There was a limited outbreak in 1956, but by this time large-scale inoculation was possible and it was quickly contained. The disease had ceased to have any special significance by the early 1960s, when the compulsory registration of births and deaths and the introduction of a hospital registration system enabled such matters to be scrutinised with some accuracy.

Tuberculosis proved a more intractable problem. Of all the ailments of the desert, this had taken the firmest hold. Philby, Thesiger, Thomas, Dickson and other Europeans who moved among the bedu frequently spoke of the cruel coughing bouts that sometimes afflicted entire Arab families. There were few encampments where at least one member of the family did not every now and again cough blood. Treatment, up to 1950, was virtually non-existent. Wards were set aside for the purpose at the Emiri hospital but they were nowhere near adequate. In 1952 the first sanatorium was opened, its 100 beds used for both men and women patients. A year later temporary accommodation was found for women sufferers, but it was not until 1959 that the state was able to deal with the problem on a sufficient scale. The massive sanatorium that was opened in that year on the seafront at Shuwaikh was a model TB treatment centre. It consisted of a three-storey, air-conditioned building surrounded by a flower garden, with every conceivable facility for treatment and recuperation. In 1956, an anti-tuberculosis campaign was begun in conjunction with the World Health Organisation. A chest consultant with four medical assistants, a research team, nurses and radiologists used a mobile unit to travel around the desert in order to find out the real extent of infection and to encourage sufferers among the bedu to submit to treatment. By keeping a close watch on affected families and by inducing many of them to go into the sanatorium, they have undoubtedly cut down the incidence of the disease, and in the towns they have largely abolished it. But even with the modern amino-salicylate drugs, which are used in vast quantities, maintaining effective control over treatment of the bedu remains a difficult task. In 1962 some 57,000 people out of a population of 321,000 were X-rayed. Of these, 1,282 were admitted to sanatoria.

The Al Sabah hospital, opened in June 1962, marked the culmination of a decade of tremendous activity and spectacular spending on the welfare services. Occupying 407,000 sq. ft, its cost in bricks and mortar alone amounted to KD 3·7 million (the conversion rate at that time was one of parity with the pound sterling, or KD 1 to 2·8 US dollars). A further million dinars was spent on medical and surgical equipment and ancillaries, including a helicopter landing pad to bring in patients from the desert and island territories. For several years from the mid 1950s, the country's health budget remained at about eight million dinars. By the mid sixties it had reached KD15 million and in 1970 it exceeded 16 million.

Impressive though they may be, these figures are dwarfed by expenditure on education. Naturally enough most of the country's professional jobs were filled by foreigners in the first affluent years. In 1949 there was one qualified medical or dental practitioner to every 25,000 people. By 1970 that figure had become one in 700. In teaching, architecture, engineering, the growth of professional services was just as dramatic, and the importation of experts just as pronounced. For ten years or more, Kuwait has been in the probably unique position of having more expatriates than natives in its population.

Narrowing this gulf became one of the principal objectives of the government. Education and, for want of a better expression, Kuwaitisation, were the twin watchwords of state policy from the moment that oil royalties began to accrue. Some indication of the ground that has had to be covered in education and finding jobs for the uneducated is contained in the Kuwait Economic Survey for 1968–9:

"It is noteworthy that over 70 per cent of the Kuwaitis in the labour force are on the government payroll. Apart from the generosity of the public social services and the piping of purchasing power in the economy through land purchase at exorbitant prices, what provides the income cushion to the ordinary Kuwaiti is the congested commercial sector of the economy and the government's readiness to accept virtually any applicant for employment regardless of his qualifications. It is significant that about 87·5 per cent of Kuwaiti government officials have either no more than elementary education or are in the uneducated category."

Before Kuwait's first school – the Mubarakeyah – was opened in 1912, a few children were taught in the homes of self-styled teachers. There was no paper available so that pupils smeared clay on a block of wood and made inscriptions with a stick. At the Mubarakeyah school the headmaster wrote his own stencilled textbooks. The emphasis was on arithmetic and letter writing, though in its later period history, geography and drawing were introduced by teachers from Palestine. The school closed in 1931 when a world economic slump caused a sudden drop in the pearl market, with nearly disastrous consequences for Kuwait.

A new start was made in 1936 when a group of teachers arrived, again from Palestine, and several small educational establishments were set up. The money was provided by a special tax levied on local merchants. In 1937, the first girls attended school – 140 in number. In the same year, some 600 boys were in class. By 1954 there were forty-one schools in the state, all stemming from what was, in effect, Kuwait's first experiment in social welfare. In 1955 an educational programme covering the ground from primary to university training was formulated. Expenditure increased by massive strides. The budget went from KD 83,000 in 1946 to KD 500,000 in 1950, to KD 6 million in 1955, to KD 14 million in 1958. These figures included the cost of buildings and in 1960–1 this part of the national budget was allocated to the Ministry of Works. The Education Ministry's budget for 1962–3 was nonetheless about KD 12 million. By 1967 it had doubled to 24 million. Between 1945 and 1950 the number of teachers in government schools grew from 142 to 294. In the next decade there was a sevenfold increase to 2,255. By 1968 the number exceeded 6,400. The student population in the same periods was, in round figures, 3,600, 6,300, 45,000, 112,000. And by 1968, more than one-third of all students were female, 47,655 in total. Here, in the persuasive language of statistics, more than educational progress is exemplified. The advance of Kuwait's women to educational and social equality in so short a time has been nothing short of revolutionary.

It was an emergency programme, dictated by the most unusual of contingencies, the availability of almost unlimited financial resources. Of course, the objectives of enlightened social expenditure are seldom realised overnight. Impatience to see a tangible return on investment is as common in Kuwait as in other parts of the world. One pointer to the success of this immensely generous educational plan, however, is the fact that by 1968 Kuwaitis constituted more than 10 per cent of the country's relatively large teaching force. Although the UAR, Jordan, Palestine, Syria, Iraq and Lebanon provided the majority of teachers, Kuwait with 680 newly trained citizens could claim a valuable asset for the future. In addition, the 274 student teachers in government training colleges in 1962 had swelled in number to 2,384 by 1967. In 1968, more than 150 Kuwaitis graduated at universities in the United Kingdom, the USA, Lebanon, Syria and the UAR. All this, of course, is additional to the private schools – and, incidentally, hospitals and mechanical training centres – maintained by the Kuwait Oil Company and various foreign national groups, and twenty or so schools paid for and maintained by Kuwait in other Gulf states.

The country's own university, opened in 1964, with 400 students, had a population of 886 by 1967, and an expansion programme designed to add faculties of medicine and engineering to its existing law, arts and science departments was under way.

57. The social and economic story in comparative figures. Index 1958 = 100. After Buchanan

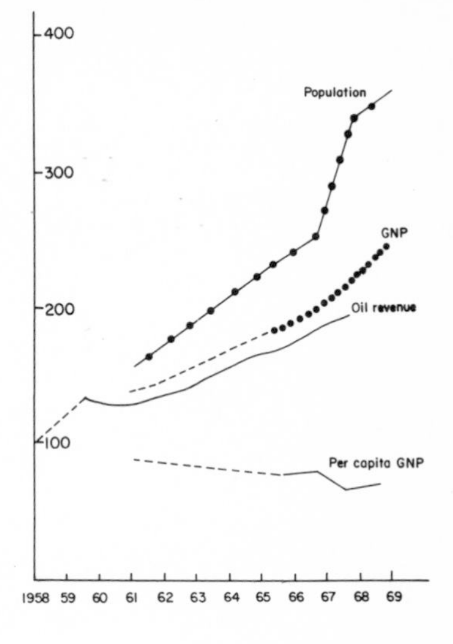

Statistics relating to Kuwait are bewildering in their magnitude and profusion. It was an ever-present problem of the oil fields that no plans, however massively or imaginatively conceived, ever measured up to the eventual requirement. This was particularly so with transit facilities and generating plant, but it applied to every aspect of development. The growth of oil output and ancillary industries continually outstripped local resources of skilled manpower and essential materials. Thus, the country set in motion a crash programme to train its own citizens to manage and maintain industrial installations, and to play a part in running social services which, if anything, developed more speedily than the productive processes that sustained them. Inevitably, though, the country had to admit foreign workers and professional advisers in numbers which soon exceeded the size of the indigenous population. The expatriate force was swelled by Palestinian refugees from the Arab-Israeli war, though almost all of these were quickly absorbed into the workforce. In eight years, from 1957 to 1965, the total population more than doubled – from 206,000 to 467,000. By 1969, the figure exceeded 700,000. The proportion of Kuwaities to non-nationals diminished from 50·4 per cent in 1961 to 37 per cent in 1969. The percentage of Kuwaitis in the labour force declined from 23 to 18 between 1965 and 1969. Another significant statistical factor in the demographic make-up of Kuwait is the high proportion of males to females. The ratio remained approximately 2:1 between 1961 and 1965, although the balance in the indigenous population is nearly even. The overall picture drawn by the population figures is one of overwhelming dependence on the educational qualifications and technical expertise of outsiders, a very slow and gradual loosening of traditional bonds which discourage Kuwaiti men from taking on certain kinds of employment and which almost rule out the idea of women working in the public and industrial sectors, and of large numbers of single expatriate workers coming in to take up the jobs that are available.

It is a state of affairs which no nation can be expected to look on with equanimity. Kuwait must and does face the fact that without the influx of foreign helpers, British, American and Arab, the country could not possibly have developed with the explosive suddenness which made it, in a single decade, one of the richest and most envied countries in the world. Equally, its government can be forgiven if it sees inherent dangers in a situation where many senior administrative jobs are in the hands of foreigners who, whatever their integrity and ability, can have no long-term commitment to Kuwait. So too in the composition of the workforce, it must look with some concern at a situation where, day by day, its own population is diluted by the foreign influx. In 1969, the ratio of employment to population was 2·7 times greater among non-Kuwaitis than among native inhabitants. The measure of the problem is conveyed by the ever-increasing rate at which foreign labour permits are issued; 32,136 in 1968 for example, compared to 14,306 the year before.

As we have seen, Kuwait is tackling the education problem with energy and resourcefulness. Its own internal expansion of primary, secondary and university training has been supplemented by generous grants to its citizens to study abroad and by equally generous aid to other Gulf regions, where several schools have been paid for and handsomely endowed. Kuwait very sensibly sees its own long-term development as being within the context of the Gulf region as a whole, rather than as that of an isolated pocket of wealth and good intention. In 1968, the country also started a broadly-based vocational training scheme in collaboration with the International Labour Office. This to some extent begins where the Kuwait Oil Company left off, with its technical training schools at Magwa and Ahmadi, by providing instruction in the basic skills of electrical and mechanical maintenance, the supervision of distillation and chemical plant, welding, and the repair of those vital manifestations of the affluent society, air-conditioners, refrigerators and television sets. A second such institute, catering for technical, administrative and management training, was due to open in 1971.

Self dependence and, as an obvious corollary, the diversification of industry, are the urgent aims of all Kuwait's planning.

In order to probe the cause of Kuwait's uniquely wealthy yet dependent social structure, however, and perhaps to seek long-term solutions to that dependence, it is necessary to look at the basic economic facts. And no economic appraisal of Kuwait can neglect to take account of the preponderant influence of oil. There is and always will be a natural imbalance in the country's economy due to the immensity of its oil reserves and to the relative ease with which its crude oil is obtained. It is an economic problem which most nations would gladly endure, but its side effects are not to be underrated.

Oil production and revenue

Production figures for Kuwait itself, the Neutral Zone (from 1954) and off-shore (from 1961) all bear out the conviction that the oil fields in and around this country are fertile beyond any initial prediction or expectation. The claim of one of the geologists of the early exploration days that he would gladly drink all the oil discovered in the Gulf, would have seemed a trifle boastful in more recent times.

Over the first twenty years, production increased every year by an average of 35 per cent, reaching an aggregate of 706 per cent by 1966. From the first year's output at Burgan of 5·9 million barrels to the 940 million barrels of 1970, the story has been one of continual, if erratic, growth. By far the largest part of the total volume comes from the KOC fields; from 100 per cent in 1953 they had dropped by less than 10 per cent to 91 per cent of all Kuwait's production in 1967, despite the combined contributions of Aminoil and the Arabian Oil Company fields. At the same time, Aminoil had serious difficulties in the Neutral Zone. Its production rose to 35·6 million barrels in 1963 from an initial 2·8 million, but by 1967 the figure was down to 24·8 million and in 1968 there was a further decline of more than 38 per cent. This was due largely to the high sulphur content of the Neutral Zone crude. By the end of 1968 the company had built a desulphurising plant designed to reduce the sulphur content from 4·2 to 1 per cent, from which it stood to gain a profitable by-product as well as a more readily saleable crude.

Although KOC's production showed some remarkable jumps over the years, mostly brought about by events such as the Iranian nationalisation, various Suez crises and the Iraq threat, the company maintained a commendable consistency in its development programme. As we have seen, one record followed another over the twenty-five years separating the first export consignment from the anniversary in 1971. In 1951, a staggering 63 per cent increase in production was recorded over the previous year. On the other hand, 1956, 1959 and 1961 saw the annual increase needle fluctuating around the 0·5 to 2 per cent mark. But looked at over this twenty-five-year period, during which KOC contributed the lion's share of Kuwait's oil production, an average annual increase of more than 30 per cent cannot be complained of. The government has, however, tended to look abroad to growth figures in such countries as Iran, Saudi Arabia, Libya and Algeria, ranging from 11 to 41 per cent, and to insist upon an acceleration of its own output figures. The KOC gave an undertaking in 1967 to endeavour to keep the annual increase at a minimum level of 6 per cent, and in order to provide for a continuing rise in the

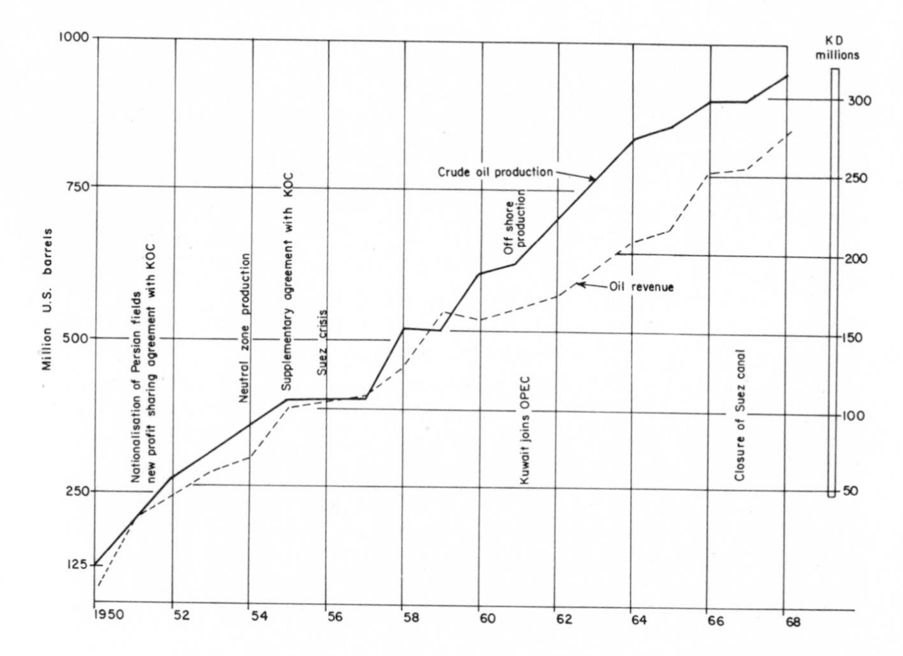

58. Oil production and revenue

level of production it formed an organisational unit known as COPE (crude oil production expansion). While accepting that there are production problems, associated with transportation and the relatively high sulphur content of Kuwait's crude (averaging 2·6 per cent), which can and are being eliminated, it may be unwise to push the growth rate of Kuwait's oil fields too hard or too fast. The real challenge is not so much to obtain more revenue from oil as to use its vast fund of natural wealth with a thought for the future, by diversifying industry and developing other resources. Admittedly, Kuwait's oil reserves are immense – an estimated 10·4 thousand million tons, which is almost one-third of the entire Middle East total, and 16 per cent of the entire eastern hemisphere's proven resources. But not all oil-producing countries are as provident as Kuwait in the use of their revenue. Many of the states of Africa and the Middle East are in dire need of money to finance military escapades of one kind or another, some are corrupt to a point where it is impossible to discover what happens to revenue from oil or any other source. Kuwait can afford to conserve its inheritance, at least within the bounds of a sensible expansion programme.

One area in which the need to conserve a valuable natural asset has been evident for a long time is the gas that for several years was flared off in the fields at both the high and low pressure ends of the separation process. With the installation of LPG plant and the subsequent introduction of reinjection facilities, KOC made a useful contribution to the arresting of this wasteful process. The development of state industries and the increasing use by state and oil companies of gas in the generation of electricity and desalination of sea water, provided further savings. In addition to the figures for KOC utilisation and reinjection already quoted, the state increased its consumption of gas from 12,391 million cu. ft in 1964 to 33,966 million cu. ft in 1968. Altogether, state and company utilisation figures for the same period increased from 72,521 million to 171,580 million cu. ft. The gas produced in the oil fields in 1968 amounted to 478,958 million cu. ft, of which 53,678 million cu. ft was reinjected. Despite this considerable improvement in the situation, more than 250 million cu. ft of gas went to waste during 1968.

Kuwait's revenue graph shows some sharp jumps and flat spots, associated with changes in the KOC concession agreement, with fluctuations in world demand, and with the international political events that have pervaded the country's post-oil history.

The production increase of 1951, no less than 63 per cent over the previous year, did not reflect with any accuracy a current profitability figure for the country's oil, since its price was still based on the 1934 royalty agreement. The following years, however, saw a dramatic change of fortune. By 1953 production was 314·6 million barrels and Kuwait's income 192 million US dollars (KD 60 million). By 1955, production reached 398·5 million barrels and income 280 million dollars (KD 100 million). Despite the Suez flat spot, which brought the annual production increase down to 0·6 and 4·7 per cent in 1956 and 1957, the country's revenue multiplied nearly three times by the turn of the decade. With a new accountancy period complicating the picture slightly by adding two months to the 1959–60 figures, a revenue of KD 167,290,000 was good enough to show that oil was contributing the lion's share to a per capita income of more than

KD 1,000. The lucrative off-shore Japanese production was also beginning to make its contribution by this time. Although the new posted-price arrangement came into effect in 1955, its real effect was not felt until the levelling of production caused by the Suez crisis had been overcome by a new spurt.

Between 1960 and the end of 1964, oil revenue jumped from approximately KD 159 million to KD 206 million, a 30 per cent increase, while production went up by a slightly higher proportion, 35 per cent in fact. In the next three years, from the beginning of 1965 to the end of 1967, revenue grew by 12 per cent from KD 216 million to KD 242 million, while production showed a rise of less than 6 per cent. The country was, therefore, benefiting significantly in income by this time, despite a levelling of the output curve. Nevertheless, the government sounded a warning note. It must be granted that Kuwait lagged behind almost every other major producer in terms of income per barrel of oil over the twenty-five years from 1946. In 1953 Kuwait was getting 61·4 US cents for a barrel of exported crude against 75·3 cents obtained by Saudi Arabia, 79·9 by Iraq and 80·1 by Venezuela. The Middle East average at this time was 70·8 cents per barrel. By 1968, the Kuwait figure stood at 81·3 against a Middle East average of 85·8. By this time, Libya was enjoying an income of one dollar seven cents a barrel. To some extent, these figures reflect the relatively high sulphur content of some of the country's oil and its high specific gravity, which help to depress market value.

One of the requirements of OPEC is that the member states should co-operate to achieve a unified production programme. The principle was established in 1965, using 1964 as the base year for subsequent comparison. A ceiling rate was fixed for each member country which, while taking account of its size and revenue requirements, would also help to combat excess production and thus an inevitable fall in price. Kuwait's growth target was fixed at that time at 10 per cent per annum, against a realised figure of 6·5 per cent. Regaei El Mallakh in *Economic Development and Regional Co-operation: Kuwait*, points out that the country is preponderantly dependent on oil income, and

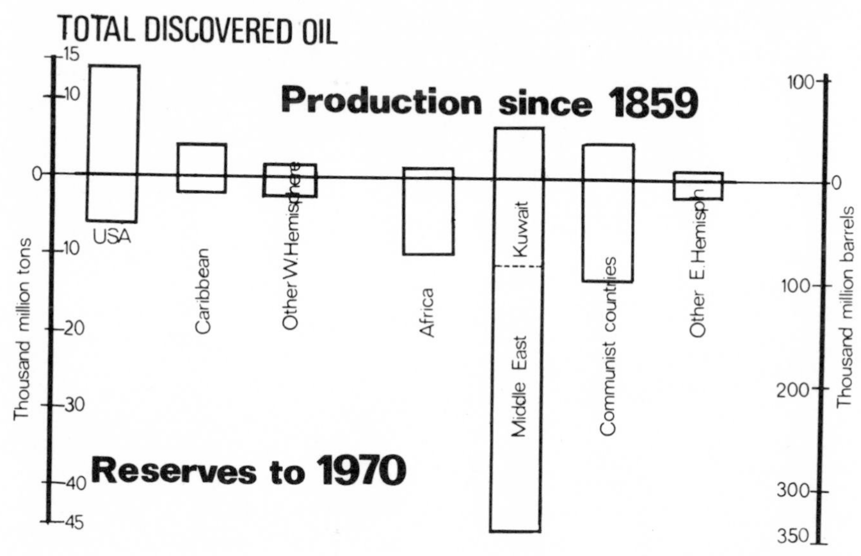

59. *Oil production and reserves in major producing areas*

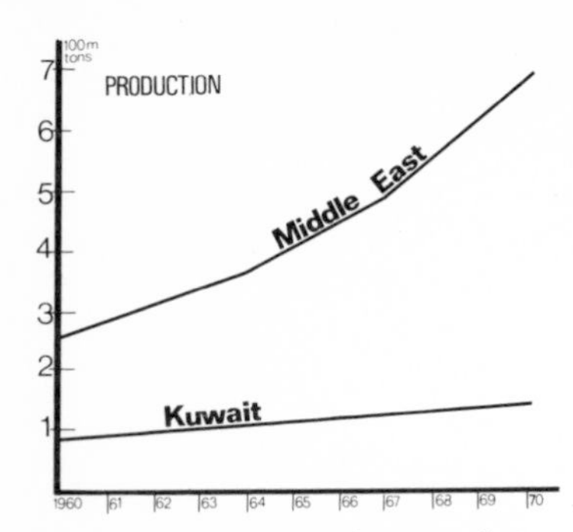

60. Kuwait's oil production during ten years 1960–1970 compared to Middle East output

that its budget is a direct reflection of oil exploitation. He also suggests that Kuwait, with its abundant reserves, would defeat its own economic purposes by pursuing a policy of excessive caution in drawing on those reserves. With an annual production rate in 1964 of 1·1 per cent of its proven reserves, compared to the United States figure of 7·4, Kuwait was in his view under-producing by a considerable margin.

"Within this context, Kuwait can ill afford any sizable cuts in its production rate. A decrease in or stagnation of governmental revenues would be felt not only domestically in the development and diversification programmes but also internationally in those areas which are recipients of Kuwaiti assistance and/or investment," writes Ragaei El Mallakh.

Be that as it may, Kuwait's is a capital surplus economy, even allowing for generous aid to fellow Arab states. While other producers fail to show the same enthusiasm for restrained advance (several important members of OPEC objected to the production programming idea from the beginning), Kuwait can afford a sceptical view of overproduction in the Middle East generally. El Mallakh's argument is double edged. As competitive energy resources gain ground it is, he says, the middle and heavy grades of oil (such as Kuwait's) rather than the lighter types, which will face stiffer competition and have to adjust to conditions of elastic demand. The cost of nuclear energy will finally set the ceiling at which fossil fuels can be sold. He quotes Adelman's *Crude Petroleum* as predicting a 28-cent nuclear price for the fuel equivalent of a barrel of oil costing 1 dollar and 74 cents. Such competition may be a good way off, but there are other ways of preparing for the battle than causing a glut in the present, and using up reserves at a dangerously rapid rate in the process.

The price factor came into prominence again in November 1970, when the posted price of Kuwait crude oil went up from 1·59 dollars to 1·68 dollars a barrel following a preliminary round of OPEC negotiations. When the talks resumed in Tehran early in 1971, a further substantial increase was called for. In the end, the agreed all-round increase sent the Kuwait posted price up to 2·085 dollars with a separately negotiated rise taking effect in June, bringing it to 2·187 dollars. Annual increases agreed at Tehran were designed to achieve a level of 2·509 dollars per barrel by 1975.

The dangers inherent in the cost-price relationship between producer and consumer countries, bringing oil into ever closer conflict with alternative fuels, was spelled out in unmistakable terms by the chairman of British Petroleum, Sir Eric Drake, in his 1971 statement to shareholders.

"The whole of the year has been a struggle to recover additional costs in our selling prices. In the early part of 1970 we had rising freight costs; now we have rising taxes and royalties in producing countries stemming from an increase in government take which the oil industry had to concede in the autumn of 1970 to producers in the Gulf and Mediterranean. . . . As a result of further demands early in 1971 from the producing countries, we in the industry came to the conclusion that the time had come when we should try and obtain, in the interests of the consuming countries, a greater degree of stability over a period of years in the price of crude oil. A meeting was accordingly arranged in Tehran in February between the oil companies and the Gulf Committee of OPEC – the Organisation of Petroleum Exporting Countries. This meeting was an important landmark in the history of the industry. The talks succeeded to the extent that an agreement was reached which is to last for five years, but the government take per barrel is now about 40 per cent higher than in October last, and by 1975 will be about 70 per cent higher. . . . Increases in price of the order agreed in Tehran can no longer be absorbed by the industry, and must therefore be passed on as quickly as possible to the consumer. This will cause a rise in the cost of energy all over the world . . . last year 53 per cent, or £1,395 million, of BP's gross income was paid in taxes of one form or another to consumer and producer governments. Of the balance of 47 per cent, we paid away 44 per cent in costs covering production, refining, distribution and transport, above all of course ocean freight, but also including depreciation and interest. Only 3 per cent remained as the net income of the group – 15 US cents per barrel or ¼p a gallon."

Diversification

Kuwait remained for the first quarter-century of its period of industrialisation in the unusual position of depending almost entirely on a single basic product. Between 1965 and 1968 the gross national product increased from KD 591 million to KD 793 million, a growth rate of 10·3 per cent per annum. It is an attractive state of affairs, but

Share of oil in Kuwait's economy (million dollars)

Year	Oil exports	Other exports	Total exports	Oil as % of total	Value of imports	Balance of trade
1958	954	15	969	98	210	+ 759
1959	863	20	883	97	260	+ 623
1960	998	23	1,021	97	241	+ 780
1961	994	25	1,020	97	249	+ 771
1962	1,120	22	1,143	98	285	+ 858
1963	1,193	29	1,222	97	323	+ 899
1964	1,316	33	1,350	97	322	+1,028
1965	1,380	39	1,419	97	377	+1,042
1966	1,428	38	1,466	97	462	+1,044
1967	1,327	36	1,363	97	593	+ 770
1968	1,484	58	1,542	95	641	+ 901

not necessarily healthy. To be able to plan in freedom from such obstacles as lack of foreign exchange, an adverse balance of payments and the incessant pressure of population increase, is as near as a nation can get to an economist's utopia. Yet these virtues can pose the serious problem of lack of incentive in other sectors of the economy. In a sense, the greater the income from oil, the less urgent becomes the need to develop other basic industries, apart perhaps from those in the service sector. In 1968, oil still represented a concentration of 96 per cent in the country's export mixture. In 1969, the position had improved by less than 1 per cent.

Thus, the emphasis in Kuwait, at least in government circles, on the need to diversify industry. In fact, the spread of manufacturing industry began almost accidentally as an offshoot of the supply units set up on the coast at Shuwaikh in the early 1950s. Water provided the first incentive. The ever-present need for fresh water in the desert gave rise to an automatic priority when money first became available for large-scale industrial expansion. When Philip Southwell arrived in Kuwait as managing director of KOC he told the Amir with obvious pride that the company intended to plant trees as the first step in a programme of providing much needed greenery. "Water before trees, Mr Southwell," replied the Shaikh, whose people had for centuries depended on the salt brackish water of the desert wells. It was a valid argument, for not much will grow without water in the Arabian desert. It was to Shuwaikh that the KOC commandeered dhows brought much-needed sweet water from the Shatt al Arab during the late forties and early fifties. In 1953 the first government water distillation plant came into operation there. Using submerged-tube evaporators it converted sea water into a million gallons of potable liquid a day, thereby supplementing the oil company's plant at Mina al Ahmadi. Brackish water was, after a while, added to this distilled water to provide essential minerals. This plant used natural gas to fire the boilers which, in turn, produced steam to heat the sea water. The same boilers were also used to drive a turbo-generator set with a capacity of 2,250 kw. Thus, in more-or-less solving the country's drinking water problem, a start had been made towards the development of a power industry.

A separate power station was built in 1954 with four 7·5 Mw generators, designed to feed steam at the correct temperature and pressure to a second distillation plant. By 1957, however, consumption had reached a point dangerously close to the reserve capacity of the Shuwaikh power installation. More generating plant was commissioned in the following eighteen months, adding 40 Mws to the state's capacity and so saving the situation temporarily. But with cables and sub-stations between the power stations and Jahra, Hawalli, Salmiya and other remote areas having to cope with an ever-increasing demand, the entire system had to be overhauled in 1958.

A third generating plant followed in 1961 when the first of three 30 Mw units was commissioned. The other two units, installed in 1962, brought the country's total capacity, excluding the oil company's supply, to 160 Mw. Yet demand soon came dangerously close to the limit again. Thus, by 1965, a new industrial expansion programme was started up at Shuaiba. This time three 65 Mw generators were installed.

From 1957 to 1962, consumption rose from 119 million to 418 million units. By 1967, the figure had reached 1·3 thousand million units. And of this total, only the relatively small amount of 117 million units was consumed by industrial plant. Here is a very practical illustration of the imbalance in Kuwait's economy created by a massive growth in domestic wellbeing, exemplified by almost universal ownership of cars, refrigerators, air-conditioners, and television, on the strength of a single product, oil.

Water usage illustrates the same point. In 1954, some 237 million gallons of potable water was produced. Ten years later the figure was 2·2 thousand million, and in 1967, it reached 4·2 thousand million. All this with the addition of another million-gallons-a-day plant in 1953, and a totally new multi-stage flash evaporator plant installed in 1957 and 1958, with four units each capable of producing half a million gallons a day. The state was thereby able to supply up to twelve million gallons a day. Yet by 1960, two further multi-stage units were required, each the largest of its kind in the world, with an output of a million gallons a day. In addition, three evaporator units were installed at Shuaiba from 1965 onward, working in conjunction with the power station. Each of these produced a million gallons a day. The sea was thus giving up seventeen million gallons of fresh water every day, largely for domestic consumption, by the late sixties. A far cry indeed from the days when the wells of the desert were important enough to be fought over, and fresh water had to be carried into the desert in goatskins filled at the town's dhow harbours.

It was, incidentally, among the minor ironies of Kuwait's development that nearly ten years after the commissioning of the first Shuwaikh distillation plant, engineers building a road from Kuwait to Basra found a sweet water well at Raudhatain, ten miles from the border with Iraq. It turned out to be a vast reservoir, twelve miles long and three miles wide, capable of producing five million gallons a day for more than twenty years. By 1962, it was being pumped into Kuwait at the rate of two million gallons a day; a welcome addition to the distilled water from Shuwaikh and Shuaiba.

These government supply industries represented the embryo of economic expansion. It took some time to assume infant shape, however.

In 1961, the government took the initiative in setting up state-backed, privately-run enterprises, with Shuaiba as the geographical location of the programme. A feasibility study was commissioned and a report prepared by 1963. As a result, the government gave substantial backing to ten industrial concerns, the most important of which, in terms of a spread of production based on existing raw material, were the Kuwait National Petroleum Company and the Kuwait Petrochemical Industries Company, out of which the Kuwait Chemical Fertiliser Company emerged. By 1965, the government had invested KD 8 million in KNPC and held 60 per cent of its equity. Additionally, a KD 23 million loan was made to the company to finance a new refinery at Shuaiba. At about the same time, the government paid KD 13 million for an 80 per cent share of the petrochemical company and joined with Gulf Oil, BP and private interests to set up KCFC. Total investment in these joint stock companies reached KD 44 million in the five years 1961–5. In addition, the Savings and Credit Bank was established in 1965 to help with further finance for industry.

Thus, oil remained – even within the diversification programme – the basis of advance. Shuaiba became the site of the world's first all-hydrogen refinery. The legislature enacted the National Industries Law, also in 1965, to give sanction and preferential terms to the participants in this costly and ambitious industrial drive. It provided for corporate income tax exemption, freedom from duty for all exported domestic products, and tariff protection for ten years. During this period, and indeed up to the seventies, these efforts produced little by way of statistical evidence to show that the imbalance in the country's internal economy was being ironed out. Starting from a base figure of KD 4 million for exports in 1953, only KD 15 million was achieved by 1965, whereas imports increased in the same period from KD 30 million to KD 134 million. In 1967–8, exports had reached an estimated value of KD 550 million, of which oil represented KD 530 million, while imports had achieved a new record level of KD 229 million. There was still a favourable trade balance of KD 321 million. Looked at another way, oil as a proportion of total exports fell by a mere 1 per cent, from 98 per cent in 1958 to 97 per cent in 1965. The figure for imports per head of population reached KD 290 annually by 1965. Government participation in the country's major industries is considerable. Apart from its holdings in the oil and petrochemical joint stock companies it had, by 1966, a 51 per cent share in the Kuwait Investment Company as well as a 98·2 per cent share in the country's foreign trade agency; in the two major transport companies, Kuwait Transport and Kuwait Navigation, it held respectively 50 and 75 per cent holdings; 51 per cent of the National Industries Company (bricks and cement) and 50 per cent of Kuwait Flour Mills; 25 per cent of the Kuwait Hotels Company. In addition, commercial banks have comparatively vast funds representing a mixture of private savings and government deposits, on which no charge is made or interest paid out. Credit on unsecured loans was fixed in 1961 at 7 per cent. In 1965–6 income from government assets abroad totalled about KD 17 million, while private foreign assets brought in KD 30 million. There is no personal income tax but there is a 50 per cent tax on income over KD 400,000, for non-Kuwaiti participation in Kuwait business. Between 1961 and 1964 more than 384 million dollars (KD 137 million) was spent on land purchase, about half of which was for private development. Expenditure on current account, primarily devoted to health, education and public works, totalled 769 million dollars (KD 274 million) during the same three-year period. Capital investment added up to 252 million dollars (KD 90 million). In looking at the other side of the balance sheet, it is significant that oil revenue, at 1,495 million dollars (KD 534 million) for the period, more than balanced the entire national budget. Overseas investment income at 122 million dollars, and revenue from public utilities, the resale of land and other sources, 131 million dollars, contributed to an aggregate budget surplus of 334 million dollars (KD 120 million), or an average surplus of over a 100 million dollars a year. It is a remarkable, enviable and in some ways untenable economic situation.

Thus, Kuwait's first five year plan.

Blueprint for the future

Five year plans are notoriously liable to fall short of their targets. It seems that human society is, on the whole, better at generating wealth than at deploying it. Perhaps, too, there is something psychologically wrong with five years. Six might give more time for readjustment, four a greater sense of urgency.

The principal objectives of Kuwait's plan, spanning the period 1967–8—1971–2, were set out by the country's planning board as: (a) to achieve a compound rate of growth of 6·5 per cent a year; (b) to diversify the economy and broaden by a relatively faster development in the non-oil sectors; (c) to develop human resources by raising standards of education, training and health; (d) to relate the educational system to the production of skills required for the development of the economy; (e) to secure a reasonable balance in the growth of different areas in the country with a view to ensuring a widespread diffusion of development benefits; and (f) to strengthen economic ties with other Arab states.

The cost of all this was estimated at KD 912 million, of which KD 507 million would be absorbed by the public sector, KD 345 million by the private sector and KD 60 million by the mixed segment of the economy.

None of the objectives are easy of realisation. In economic terms,

State Budget
Ordinary Expenditures
1965–6—1970–1
(KD '000)

	65–6	66–7	67–8	68–9	69–70	70–1
Head of State	10,000	8,000	8,000	8,000	8,000	8,000
Court of Shaikh	111	208	231	224	226	230
State auditors	492	637	542	517	432	417
Cabinet	212	240	277	482	511	523
"Fatwa"	116	147	154	155	148	153
Planning Board	348	426	396	533	712	696
Adjudication commissions	—	96	101	104	105	110
Civil servants	459	505	545	481	492	504
Complementary allocations	—	—	—	1,237	646	2,937
Public works	9,105	10,033	10,541	10,135	10,785	10,679
Guidance and information	6,118	4,863	4,988	4,812	5,140	5,191
"Awqafs" and Muslim affairs	786	1,027	1,166	1,158	1,274	1,332
Telephone and telegraph	2,477	2,414	2,555	2,621	3,097	3,233
Post	865	982	1,083	1,065	1,160	1,201
Commerce and industry	305	513	483	506	545	625
Education	16,751	19,836	24,834	27,390	30,354	31,413
Foreign affairs	2,474	2,753	2,835	2,896	2,967	3,064
Interior	13,188	15,219	17,118	17,866	20,161	20,939
Civil aviation	420	770	868	1,002	1,030	1,028
Defence	11,000	13,000	21,500	23,000	25,000	25,000
National Guard	—	—	—	1,000	1,000	1,000
Social affairs and labour	3,716	4,328	4,434	4,690	5,563	5,806
Public health	11,479	13,541	15,133	15,724	16,364	16,608
Justice	1,028	1,181	1,470	1,502	1,515	1,584
Water and electricity	6,169	6,946	7,704	7,631	7,969	8,237
Shuaikh stations	878	965	1,079	1,019	1,022	1,060
Chlorine plant	106	133	160	145	174	204
Shuaiba stations	—	373	453	457	549	898
Finance and petroleum	892	1,575	1,405	1,279	1,212	1,178
Customs and ports	4,705	5,118	5,919	6,443	6,040	6,117
Housing	2,437	2,623	2,771	2,631	2,799	2,852
International airport	—	—	—	167	—	—
Miscellaneous and transfers	29,593	33,116	54,663	49,595	75,027	79,102
TOTAL EXPENDITURES	136,230	151,568	193,408	196,467	232,019	241,921

Source: Kuwait Chamber of Commerce and Industry Magazine.

the scarcity of locally available skills and raw materials imposes disciplines on growth and diversity whatever plans may be formulated. Apart from this fundamental difficulty, an unforeseen setback occurred in the initial stage. The plan had received the clearance of a special committee of the National Assembly but was not government approved by the middle stage of its first year. Then, in June 1967, the Arab war with Israel made a sudden additional call on Kuwait's resources. It is difficult to assess the exact cost to Kuwait of the interminable conflict caused by the Israeli occupation of Palestine, but an indication may be given by a comparison of the overall contributions to the Kuwait Fund for Arab Economic Development, an essentially businesslike body which looks at every request for aid in conservative terms, and separate loans put down to the UAR and Jordan. In 1968–9 KFAED loans to the Arab world amounted to KD 68,810,000 while an extra KD 47,140,000 was allocated to the two countries most affected by the war effort.

As a result of this additional expenditure in the first two years, results fell below planned expectations, the public sector predictably taking the brunt of the shortfall. The report of the planning board went so far as to say that in its first year the Five Year Plan was a non-starter. At any rate, in the first two years, only 19·4 per cent of the projected outlay in the public sector had been spent. It was hoped that by achieving a higher rate of implementation for the third year, 1969–70, a 28 per cent realisation of the first three years' proposals would be reached.

Estimated employment of national resources
1966–7 and 1971–2
(KD million)

	1966–7		1971–2	
Gross national product*		604		842
Consumption				
Private expenditure	210		309	
Public expenditure	120		161	
		330		470
Investment (capital formation)				
Private sector	73		81	
Public sector	63		127	
		136		208
Changes in stocks		14		—
Balance of payments surplus		124		164
TOTAL		604		842

*From plan estimates

While investment in almost every part of the public sector was disappointing during the first three years of the plan, activity in the private and mixed areas exceeded the planning board's hopes. In these sectors KNPC made significant progress in its all-hydrogen Shuaiba refinery, while the Kuwait Fertiliser Company, after a hesitant start, achieved production records in 1967 and 1968. New plant was under construction or at the "drawing board" stage for all the major industries at Shuaiba throughout the period of the Five Year Plan. Investment in the private and semi-public sectors was of the order of KD 118 million for 1967–8, KD 114 million for 1968–9, thus keeping pace with the programme at 57 per cent of the original budget.

The paradox of the position is made very clear by the basic facts. Between 1965 and 1969, the gross national product grew from KD 591 million to KD 793 million. By the latter year it had reached KD 1,133 per head of population. That population consisted of 259,000 Kuwaitis and 441,000 non-Kuwaitis.

Estimated gross domestic and national products base year 1966–7—target year 1971–2 (KD million)

	1966–7		1971–2	
Gross domestic product at factor cost		801		1098
Less: Net factor payments to rest of world	200		260	
Gross national product at market prices		601		838
Plus: Net indirect taxes and subsidies		3		4
Gross national product at factor cost		604		842

These figures do not take into account depreciation of fixed capital assets, assumed by planning board to be about 10 per cent per annum, on a base figure for fixed assets of KD 800 million and an estimated target of KD 1,220. Thus, figures for gross national income for the two periods are about KD 524 and KD 716 million.

Overall growth, despite the disappointment of low investment in the public sector, was higher than anticipated. In the first two years of the plan the GNP reached a compound growth rate of 16 per cent against an estimated 14·3 per cent. Nevertheless, there was a continued expansion in immigration and the foreign labour force. The Shuaiba refinery, for example, had a workforce of 831 at the end of 1968 made up of thirty-one nationalities. Failure to meet educational investment targets only exacerbates that kind of situation.

According to the *Arab Economist* of March 1971, "The Kuwait economy was still in 1970 under the effect of the stagnation which characterised the activity of most of its sectors following the 1967 Arab-Israeli war." The journal added, with some truth, "The basic problem lay in the fact that the huge oil revenues were reinvested haphazardly in an unco-ordinated way with the result that some sectors (construction, commerce and services) were favoured to the detriment of the others, creating basic disequilibria in the economy. To these structural factors were added, from 1967, other exogenous factors such as the annual subsidies that Kuwait promised to grant to UAR and Jordan at the Khartoum Summit Conference, the rise in interest rates in Europe and the slow down in the rate of growth of oil production." Such strictures, coming as they do from an Arab source, may carry weight in Kuwait, where the business community, always powerful as a lobby, has found its political feet in the new National Assembly. By 1970, these people were voicing their views loud and clear. They went so far, in fact, as to turn down a joint enterprise agreement with the Kuwait Oil Company which had received ministerial blessing, on the grounds that it would be more profitable to go it alone. But when all is said and done, Kuwait has shown more purposefulness and more adaptability to the wind of economic change than any of its neighbours. Those neighbours should not expect too much of their more provident brethren.

61. Sir Gawain Bell, one of the last and most popular of Britain's Political Agents in Kuwait. He was there from 1955 to 1957 before becoming Governor of Northern Nigeria and was succeeded by Mr A. S. Halford

Political superstructure

Treaty of Independence is the term often used to define the act by which, in 1961, Kuwait assumed full control of its own processes of government and its equality of partnership in the councils of the world. We should be careful about the use of this essentially loose definition, however. Kuwait was never colonised either by Turkey or Britain. It never ceased to be a sovereign state in any legal sense. The fact that Turkey had nominal suzerainty over virtually the whole of the Arabian peninsula up to the 1914–18 war does not materially affect the issue.

In order to see the position of Kuwait and other Arab states in the perspective of imperial geography, it is necessary to turn again to events after the first world war which marked the break-up of the Ottoman Empire and the ascendancy of Britain and France. The Arab revolt against western cultural and commercial influence had become an irresistible force by the time the allied powers, with internal assistance, had driven the Turks from the peninsula. It did not take the imperial powers long, however, to confuse a fundamentally simple picture. On 2 November 1917, A. J. Balfour made his famous declaration to the Zionists promising British support for the establishment of a Jewish national home in Palestine. A month later the Bolsheviks caused the secret Sykes-Picot agreement of 1916 to be published. This partitioned Syria and Mesopotamia into a French-controlled sphere in the north while the British were to govern the south. Palestine was to be administered internationally. The seeds of dissension were sown even before the armistice. With the conquest of Palestine and Syria by Allenby, the Anglo-French statement tried to clarify the position:

"The goal envisaged by France and Great Britain . . . is the complete and final liberation of the peoples who have for so long been oppressed by the Turks, and the setting up of national governments and administrations that shall derive their authority from the free exercise of the initiative and choice of the indigenous populations."

Naturally enough, the Arab world took this as a specific pledge of autonomy. In 1920, that hope was dispelled by the San Remo Conference at which the Allied Supreme Council agreed that all the states of the peninsula should become independent, except for the Mediterranean countries. The latter were to become mandated territories under Britain or France. Hejaz and Najd became independent kingdoms, and Britain guaranteed the security of smaller states, such as Kuwait, which might be threatened by the emergence of these great powers of the interior. The last vestige of claim to authority of any kind over these territories of central Arabia and the Gulf disappeared in 1924 when Mustapha Kemal, president of the new Turkish Republic, formally abolished the Ottoman caliphate in Islam.

One country has to occupy another and control at least its basic military and political affairs in order to stake a claim of responsibility. Britain gave a promise of protection in return for services rendered before and during the first world war, as it did to other shaikhdoms in the area. Equally, Kuwait made assurances to Britain which gave priority to its citizens in specific fields, such as exploration for oil. Britain, through the government of India, directed the country's foreign affairs and provided an administrative adviser in the shape of the Political Agent. None of this is inconsistent with the fact that Kuwait remained for the best part of two hundred years a free Arab state, able in the last resort to make up its own mind on issues of domestic policy. If evidence of this was ever lacking, it was surely made good by the insistence of the British during the negotiations for the original oil concession. The Indian government, the Colonial Office and the Foreign Office all maintained explicitly that the Shaikh of Kuwait was an independent ruler who could be guided by Britain, but no more. Thus, when foreign writers and Kuwaitis themselves talk of independence they employ an inexact term.

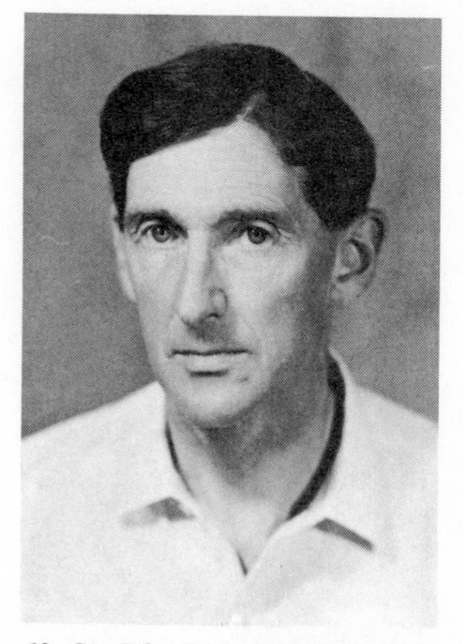

62. Sir John Richmond, last of the Political Agents and Britain's first Ambassador

Nevertheless, the treaty of 1961 had a profound effect on Kuwait's internal structure and its international posture. There were inevitable formalities. A full Ambassador was appointed in place of the British Political Agent – Mr J. C. Richmond proceeding from one post to the other – and a new flash and emblem were adopted. The emblem was an adaptation of the one instituted in 1956 by Shaikh Abdullah al Salim, a dhow in full sail encircled by a falcon and containing the proclamation: *Dowlat al Kuwait*, "Kuwait independent". The national flag of 1961, made up of green, white and red stripes in association with a black trapezoid, replaced the earlier inscribed white-on-red ensign.

Not long before the oil concession was signed, Shaikh Ahmad confided to Colonel Dickson his feelings about the advent of wealth. "You must realise," the Political Agent told him, "that this will probably mean that Kuwait will become very wealthy; a large number of outsiders will have to come to your country to work the oil; it will bring both good and bad . . . it will mean the building of a power station for lighting and air conditioning, modern building materials for houses. Think very carefully about these things." Shaikh Ahmad did not reply at once. But a few days later he told Dickson: "I must do this for my people, even if it will bring undesirable things to my country. We are poor, pearling is not what it used to be, so I must sign."

Shaikh Ahmad was the last of the traditional rulers of Kuwait. He accepted the need to develop the country's oil resources, and he accepted that change would inevitably follow. He regretted that some of the old customs of brotherhood and tribal leadership which grew out of the very austerity of a desert existence would be lost in a surge of technology and social sophistication. Shaikh Abdullah al Salim bridged the gap from a rigidly tribal society to an industrial democracy, in which the power of the hereditary shaikhdom was no longer total. His successor as Amir of the state, Shaikh Sabah al Salim al Sabah, led it into its present era of superabundance and creeping technocracy.

The 1961 treaty substituted the spirit of friendly co-operation for contractual military support in the relationship with Britain. As soon as it was signed Shaikh Abdullah set out a programme which included the immediate establishment of a provisional government through an appointed constituent assembly. This assembly was to draft a constitution for the state within one year. The draft constitution was approved by the Amir on 11 November 1962. The first general election was held on 23 January 1963. The country was divided into ten regions, each member of the electorate having five votes. In some areas as many as thirty candidates competed under this system of proportional representation, those who failed to gain a tenth of the votes losing their deposit of KD 50. Altogether, forty-nine deputies were elected, some regions returning more than one candidate.

63. Heir Apparent and Prime Minister, Shaikh Jabir al Ahmad al Jabir

The idea that Kuwait's parliament was a rubber stamp body, giving a superficial gloss of democracy to the perpetuation of absolute hereditary rule, was quickly dispelled. One of the first acts of the new National Assembly was to question the size of the privy purse. In April 1963, twelve deputies proposed that the defence arrangement with Britain should be abrogated so that Kuwait could join with Syria and Egypt in President Nasser's UAR. There was little of the cipher about this parliament. In the executive functions of government, Kuwait has somehow managed to combine American and British procedural features. There is a President whose function is much the same as the Speaker's in the British Parliament. The Prime Minister is appointed by the Amir. Of fifteen ministers, only one is elected, but all are answerable to the Assembly and a vote of no-confidence can force their resignations. The Council of Ministers can be compelled to submit its policies to the scrutiny of the Assembly. General elections are at four-year intervals.

The constitution of Kuwait, drafted by a committee of five members of the Assembly, is in many ways a remarkable document. After laying down some obvious principles, such as the sovereignty and Arabic character of the state, the pre-eminence of Islam and Islamic law, the priority of Arabic as the official language, and the obligatory use of flag, emblems, national anthem and other symbols of nationhood, it goes on to set out principles of action and behaviour such as many a country has taken several hundred years to evolve; and not always with so much conviction. Personal liberty, the freedom of the press are guaranteed, and the care and protection of the young and old are the subject of specific requirements under a section labelled "Basic Constituents of Kuwait Society". Discrimination on grounds of race, social origin, language or religion invites severe penalties. Freedom to form or join trade unions and other associations, and to contract out of them, is another constitutional guarantee, as is the right to assemble without notification or approval. Police are forbidden to attend private meetings. Political refugees cannot be extradited. The Amir alone enjoys constitutional immunity. As Commander-in-Chief of the armed forces he can declare war by decree, but only for defensive ends. The constitution specifically forbids participation in offensive warfare. Most importantly, these freedoms and limitations on the executive power, are watched over by an independent judiciary.

Historical precedents are usually questionable, but few nations can in the nature of things have jumped the wide gulf which separates hereditary tribal rule from universal suffrage with as much alacrity. It was an act of voluntary and enthusiastic faith on the part of successive rulers. But there was, of course, more to it. Political structures grow out of economic circumstance. It would have been as impossible for Kuwait to embrace suddenly a highly industrialised economy while administering its affairs and deciding its long-term policies through a small, closely-knit governing family, as for Britain to bring its industrial revolution to completion within the confines of aristocratic privilege and the divine right of kings. The magnitude of the change in Kuwait's political and social institutions was a reflection of the immense change in the country's economic fortunes. For all its inevitability, the metamorphosis was nonetheless remarkable.

Territorial integrity

Call it what we will, the arrangement by which Kuwait assumed total control over its own destiny in 1961 was soon put to the test. The Iraqi claim to Kuwait made in 1938, based on the imposed provisions of Ottoman rule, echoed across two decades.

By 1954 Kuwait had become a hot property, and Iraq's leader Nuri al Said was instrumental in that year in inviting it to join the Baghdad Pact alliance. Britain, in one of her less perceptive moods, encouraged Shaikh Abdullah to join the pact and thus solve once and for all his country's fresh water problem, while finalising a frontier situation which had remained in a state of uneasy suspension since Uqair. The ruler would have none of it, however, since he knew full well the extent of Nuri's unpopularity in Iraq and of the instability of his regime. He had been asked to trade the island of Warbah, opposite Bubiyan, in return for unhindered access to the Shatt al Arab. Thus to place a vital sea route at risk seemed too high a price for membership of an ill-assorted alliance and access to water, however tempting the latter offer may have been to a country that had for so long in the past had to rely on the brackish water of the desert. In any event, the Shatt al Arab had lost a lot of its significance by the time the Iraqis were ready to use it as a bargaining counter.

In July 1958, the Hashemite regime was finally overthrown and General Kassem came to power. Britain, regarding Kassem as unstable, and characteristically suspicious of communist influence, began to reinforce its military presence in the Gulf. One of Kassem's first acts on his assumption of power was, in fact, to borrow £66 million from the USSR in order to build a port at Umm Qasr no more than a mile from Kuwait's northern boundary. Several views have been expressed as to the facts and fictions which surround the subsequent story of a threatened Iraqi invasion of Kuwait. We will stay as close to the facts as is possible in a saga which gave rise to manifest hysteria in press comment and may well have been the result of an act of sheer bluff on the part of its instigators.

In 1958, Shaikh Abdullah went to Baghdad to discuss relations between Kuwait and Iraq with the new prime minister. No decisions were arrived at, neither were any claims made on each other's possessions or plans. A year later, in May 1959, a "plot" to infiltrate Kuwait was uncovered and two hundred Iraqis found camping in the desert were arrested. They were moved to a compound on what Violet Dickson has since described as "a quagmire". They were apparently uncomplaining and gradually resumed their meagre existence.

Meanwhile, Iraq built up its border force facing Kuwait. On 19 June 1961, following the new agreement with Britain, the latter's offer of assistance was quickly put to the test. Six days after the agreement was signed, Kassem called a press conference at which he declared that Kuwait was an inseparable part of Iraq and that he would demand every inch of its territory. According to *The Times* of 26 June, he went on: "Tomorrow Iraq would inform all nations of her position with regard to the territory." He rejected the new Anglo-Kuwait agreement and said it was a "specially dangerous blow" against the integrity and independence of Iraq and the people of Kuwait. From the date of the abrogation of the "illegal 1899 agreement" Iraq had decided that nobody inside or outside Kuwait, of whatever status should rule Kuwait, he was reported as saying. After denouncing the 1899 agreement between Kuwait and Britain, he went on: "We would stand up to imperialism if it intervened in her (Kuwait's) peaceful movement."

General Kassem said the present ruler would be warned of transgressing against the people of Kuwait, who were Iraqis, or else he would be dealt with most severely as a "mutineer". He hoped that the ruler would not obstruct the progress or the welfare of the people of Iraq.

On 27 June *The Times* reported a Kuwait radio broadcast under the heading "Kuwait to defend her Independence". It said that the government was fully confident that all friendly and peace-loving countries "especially sisterly Arab states", would support it. "Kuwait," said the announcement, "is an independent Arab state with full sovereignty, recognised internationally . . . Kuwait government . . . is determined to protect and defend its territory."

In Baghdad a memorandum was handed to foreign diplomatic missions declaring "Kuwait is part of Iraq". It went on to say "There is no doubt Kuwait is a part of Iraq. This fact is attested by history and no good purpose is served by imperialism denying or distorting it."

On the same day, 27 June, King Saud sent a message to the ruler. It said "We are with you." The King described General Kassem's stand as "strange" and added: "As far as we are concerned we are with you in your fight and struggle and stand by what we have undertaken. We are fully prepared to meet any danger to which fraternal Kuwait may be subjected."

In Cairo, the Arab League, of which Kuwait had become the eleventh full member, described the Iraq claim as a "surprise". The Shah of Persia took the opportunity to send Kuwait's ruler a message of congratulation on the attainment of independence, and expressing a willingness to exchange diplomatic missions. A *Times* report from Beirut suggested that Kassem's slighting references had . . . flung down the gauntlet to Britain as well as Kuwait.

By this time everyone who could remotely be expected to have an opinion on the subject was giving vent to it in the press and on radio and television, in every part of the world. Perhaps the most intelligent contemporary piece of advice was that the ruler should keep to his plan to spend the summer in Beirut, thus indicating that he at least did not exaggerate the Iraqi threat.

Tuesday 27 June also saw a meeting between the ruler and the Political Agent John Richmond, shortly to be appointed Britain's first full ambassador. Colonel Nasser joined other leaders of Arab states in expressing cordial greetings to Kuwait. Iraq did not seem to be overburdened with friends. Britain's obligation to assist Kuwait was stressed by the Lord Privy Seal, Mr Edward Heath. By Wednesday 28 June, the world press was reporting wide support for Kuwait. The United Arab Republic issued a carefully worded statement rejecting the "logic of annexation" among Arabs. British warships were reported to be heading for the Gulf on 30 June, and the Admiralty confirmed that units of the fleet were on the way from Hong Kong and Singapore. Following a visit by the Saudi Arabian army chief, Saudi troops were reported to have arrived in Kuwait and to have closed the border with Iraq. Iranian launches, carrying supplies to Kuwait, were fired on by Iraq patrol boats. The British government continued to underline its readiness to assist Kuwait, and by 1 July its tone had become graver: "HM Government have informed a number of friendly governments in Middle East and elsewhere of its deep concern at the situation and expressed to them the hope that they will use their moderating influence with the Iraq government, so that Kuwait may continue her development as an independent Arab state amongst nations of the world."

The Times commented: "It seems that Kassem, having staked out a strong claim to a share of Kuwait's resources, had not intended to put pressure on its Arab neighbour to the point of military attack."

The House of Commons met to hear a report on the situation. It was told that no British force was yet in Kuwait.

On 3 July a detachment of Royal Marines landed from HMS *Bulwark*. US destroyers were preparing to evacuate American personnel. More British troops with tanks and armoured cars were variously reported to be on the way or to have arrived. Arrangements for the evacuation of British personnel were put in hand.

In the course of this international emergency, Kuwait and Shaikh Abdullah remained remarkably calm. At a press conference, the ruler asked reporters of local newspapers not to provoke or irritate General Kassem in any way. He also said that he did not intend to seek a military treaty with Britain.

Announcing the landing of British troops to the House of Commons on 4 July, Prime Minister Macmillan said: "The government earnestly hopes that counsels of moderation will prevail in Baghdad. Our forces are there purely for defensive purposes, in accordance with our treaty obligations. They will be withdrawn as soon as the ruler considers Kuwait is no longer threatened."

On 6 July, Iraq was reported to be increasing its border force while British troops started to lay minefields across the northern Kuwait desert. Cairo and Moscow called for the withdrawal of foreign troops, and the commander of the British 24th Infantry Brigade said that the possibility of attack had grown. King Saud proposed a meeting with

General Kassem which the latter declined. By 7 July, with 6,000 British troops in Kuwait, the airlift into the new and as yet unused civil airport south-west of the town, slackened. On that day, Britain was instrumental in putting before the Security Council a resolution condemning the Iraqi threat. The next day, 8 July, Britain began to withdraw its troops. The Security Council reached inevitable deadlock on the resolution and by 10 July, Britain had announced that failure to agree on international action to protect Kuwait would result in a prolonged British presence. Kuwait then proposed to the Arab League that it should provide a contingent to replace British forces following their withdrawal. By 14 July, one-third of the British force was on its way out.

On 15 July General Kassem made a speech to mark the anniversary of the 1958 revolution. He renewed his claim to Kuwait and demanded the withdrawal of all British forces. "If Iraq had chosen to use force, she could have taken Kuwait long ago," he said.

He may well have been right. But if that was his view all along, he had certainly succeeded in demonstrating that the game of bluff can cause a mighty stir in the region of the Gulf.

Perhaps the best retrospective comment on this storm in a teacup was provided by David Holden, the *Sunday Times'* Middle East expert, in his book *Farewell to Arabia*.

"For sixty-two years . . . no significant British land force had ever been summoned to its [Kuwait's] defence. Now, in the first month of its independence, nearly 6,000 British troops poured into the country to prove that Britain was still in earnest about upholding peace in the Gulf. It was a traumatic experience for everyone concerned: testing for the British, whose resources proved scarcely equal to the task; chastening for General Kassem, who was scared off; humiliating for most of the other Arab states who saw – not for the first time – an imperial power settling their domestic problems for them; disillusioning for the Kuwaitis, who had hoped to celebrate their independence in a burst of Arab joy and brotherhood; and profoundly frightening for all the other Gulf shaikhdoms. . . . Most discouraging of all, perhaps, was the inept performance of the Arab League in this sorry little crisis. Divided in their counsels, as usual, the League members refused at first to help the ruler of Kuwait, who had turned to them even before he invoked his defence agreement with Britain. Then, although shamed into action by the British, and genuinely disturbed by Kassem's unilateral attempt to seize the richest of all Arab territories, the League took several weeks to produce a force of its own to replace the British troops. When the force arrived at last, nearly three months after the original Iraqi threat, it suffered an immediate and characteristic body blow through the secession of Syria from the United Arab Republic; and before six months were out it had collapsed, as first one and then another national unit was withdrawn for pressing domestic reasons. After this it was hardly surprising that the ruler withstood persistent pressure from both inside and outside Kuwait to abrogate the British defence agreement in favour of an Arab pact: his Arab brothers, alas, had left him little option by their quarrelling and incompetence."

Arms and foreign policy

Defence, along with the para-military effort involved in supporting the united Arab cause in Palestine, has made heavy calls on the country's resources. Because Kuwait does not possess a light engineering industry, much less industries of the kind needed to produce the refinements of a modern military posture, everything has had to be bought with foreign exchange and without concurrent investment in the domestic economy. The strain is far from unbearable, of course, to a country with a perpetual trading surplus, but it is a total financial loss, and like the arms programmes of all but the most technically developed countries it inhibits progress in other sectors of the economy. Prior to the Iraqi threat of 1961, Kuwait depended largely on Britain and its fellow Arab League members for such protection as was needed by a state whose policies were essentially peaceful and non-aligned in terms of power blocks and military ambition. In fact, a ministry of defence was not formed until 1962. By 1964, however, it was receiving KD 10·5 million, or nearly 10 per cent of the public expenditure estimate. The following year it was allocated a slightly lower proportion of the annual estimates at KD 11 million. In 1966 budgeted defence costs rose to KD 13 million, in 1967 to KD 21·5 million, in 1968 to KD 23 millions. From then on the annual defence estimate was consistently higher than 10 per cent of the annual domestic budget under the Five Year Plan; 23 million dinars in 1968 (17 per cent), 25 million in 1969 (11 per cent).

The army, commanded by the young Sandhurst-trained Brigadier Mubarak al Abdullah al Jabir, is equipped with the latest tanks and land warfare equipment. There is only one criterion: it must be the latest and the best. The same applies to the jets and helicopters used by the air force. Military practice and awareness became prior requirements after 1961. Manoeuvres of one kind or another, as well as tactical training for officers, are in almost continuous progress. The purpose is without question defensive, but it would become formidably offensive if Kuwait was called upon to defend itself in the 1970s and beyond.

A well-armed defensive military position was reinforced by a policy of neutrality and non-alignment in foreign affairs from the moment Kuwait took hold of the reins. Loans of great generosity to neighbouring Gulf states, to practically all Arab countries through the Kuwait Fund for Arab Economic Development, and to Jordan and the UAR as the countries most affected by war with Israel, represent one arm of a foreign policy that appears to have three major objectives. The first, Arab brotherhood, is no more than the expression of a deeply ingrained sense of solidarity with people of the same ethnic, cultural and religious background. The second objective is to steer clear of all the pervading ideological struggles between East and West, between the capitalist and communist halves of the world. The third is to establish for itself a role in the Near East similar to that of Switzerland in Europe, a kind of political Red Cross which will try to cool the tempers of friends or foes in conflict, and provide refuge and first-aid when councils of moderation have failed.

All three aims demand a good deal of perseverance. Even regional co-operation has its hazards. Neighbours may be brothers under the skin, but they can appear in strange disguises. Envy is easily excited. Help for a less well-off relative can often nourish greed. It is to be hoped that nobody will be foolish enough to confuse motives of generosity with those of weakness. As for the East–West power conflict, the peoples of the Islamic world have always tended to look on the moral and political differences of the outsider with good humoured scepticism. They see the various Western interpretations of God-fearing citizenship and communist notions of materialism in much the same light of misplaced innocence. Kuwait has opened its doors to both, maintaining diplomatic relations with any who are willing to accept its live-and-let-live philosophy. But again, economics and politics are inevitably, and often dangerously, related. The adoption of modern processes of production have brought the country face to face with the substance of ideologies that were for a long time irrelevant. Technically, the country remains heavily dependent on American and British know-how. Commercially, it is contractually tied to two giants of the capitalist world. Internally, the conflict between private and state interests in a mixed economy must be a latent cause of friction. At the start of the seventies there was talk of joint Russo-Kuwaiti development schemes. It is all in the spirit of a free but heavily subsidised industry and of the neutrality which is at the root of the state's foreign policy. And certainly nobody but a supreme optimist would look on Kuwait as a fertile source of economic discontent, or particularly good ground for sowing the seeds of subversion. A country that guarantees the rights of trade unions yet which pays such high wages that it finds difficulty in getting anyone to join them, may just prove in the end that harmony is possible within an intensive, mixed economy and in a divided world. Membership of GATT and the UN gives Kuwait the opportunity to air its principles of commercial and political pliability on an international stage that is usually doctrinaire and almost always rigidly committed, East or West, left or right.

Curiously, it was the Soviet Union that vetoed Kuwait's first application to become a member of the United Nations in 1962. The objection was short lived, however. Morocco, as an Arab member of the Security Council, made a further appeal, and on the proposal of Tunisia membership was granted on 14 May 1963.

One other latent source of friction was toned down, if not entirely removed, by the 1961 treaty. The Neutral Zone which Sir Percy Cox had, to do him justice, seen as the surest way of preventing a Saudi attack on Kuwait, ceased to have any contractual significance. Kuwait and Saudi Arabia simply drew a dotted line across the Zone, between the Aminoil and Getty concessions, and the oil companies refrained from mentioning the *de facto* division. To the outside world, however, Neutral Zone remains a convenient term in expressing the activities of the different oil regions.

9. Looking to the Future

The Kuwait that began to take premeditated shape under the Five Year Plan was a very different place from the desert region with its compact coastal township of mud-brick dwellings into which oil first spurted in 1936.

The intervening years, from the export of the initial oil cargo to the development of an industrial society, was talked and written about by the outsider with more insensitivity than was pleasing either to Kuwaitis or to Europeans who knew the country intimately. Some came in search of extravagance, overt luxury of the gold-plated Cadillac variety. Some wove a misconceived tale of romanticism which resident Europeans thought of as the "Fry's Turkish Delight" approach. Few saw in this newly rich state a uniquely generous experiment in the application of vast, unaccustomed wealth to the social good; neither did they see the upheaval in modes of thought and behaviour that such a policy must entail for a traditionally Arab nation imbued with the culture and discipline of Islam. The change was very evident to those who knew the place well.

It is doubtful whether, thirty or more years later, the paternalistic Shaikh Ahmad would have recognised the democracy that his signing of the oil concession had jerked into being. The hunting trips which were to him the highest pleasure that a life of total simplicity could offer, the sport of desert princes and their pampered hawks, had faded, like the game itself, into a dim corner of the past. The wild flowers which Violet Dickson so lovingly and meticulously collected and sent to Kew to be docketed with their Latin names, *arfaj* and *abal* and *arta* and *busal mo*, and the purple *khazama* which they called *Horwoodia dicksoniae* because the botanists had not seen it until she sent them a sample; all are scarce now. Wild animal life of every kind is nearly extinct, though the scientists insist that the last one or two are seldom eliminated, and so here and there a couple of ostriches or oryx or wolves may mate and surprise the urbanised inhabitants of the place with their progeny. But it is unlikely. Even the insects have gone, with DDT, and with their disappearance the migratory birds have taken to other routes. Only the jerboa, the hardy desert rat, the lizard and the snake seem to have survived the holocaust, and they in diminishing numbers. For older Kuwaitis, and for the very few Europeans who knew what it was like before and after, affluent Kuwait is not an unmixed blessing. The same may, of course, be said of almost any country. It is just that Kuwait's transformation was quicker and therefore in starker contrast.

Modern Kuwait did not simply knock down the old and replace it with a marble and concrete urban jungle. It looked ahead as best it could to amenity as well as to facility. It began, in the early 1950s, an ambitious programme of tree planting to break the desert winds and sandstorms, to assist the attempt to stem erosion and irrigate the more hopeful regions. Hydroponic and aeroponic methods of plant growing have produced good crops of fruit and vegetables and given a splash of colour to the forecourts and gardens of the university and public buildings. Progress, though, has often been a matter of two steps forward and one back. The priorities became obvious when building work began in earnest. Contractors denuded the seashore of its sand and shingle in order to lay the foundations and mix the concrete for houses, cinemas, civic buildings, schools, hospitals, sports arenas, roads, garages and all the other manifestations of the wealthy and socially conscious society. The intention was admirable, the speed breathtaking. The effect, when the time came to take stock, was inevitably less than perfect.

Before looking at what might have been, or might yet be, it is worth taking a look at another batch of statistics.

By 1963, after just one decade of intensive building, Kuwait had some 10,000 houses, ranging from luxury apartments to specially built lower-income establishments. There were 7,043 shops and 2,864 factories and workshops. Domestic building was based on a state mortgage scheme which enabled Kuwaitis to purchase homes without deposit at 2 per cent interest over twenty-five years. Almost equally generous terms were available to local building contractors who needed finance. Thus, there was a two-way incentive to build fast and in plenty. Between 1964 and 1967, a further 8,308 lower-income houses were built, while the sea front was ribboned with expensive blocks of flats and rich dwellings of almost every conceivable architectural mannerism. In a state with a total population of one of the larger London or New York suburbs, there were seven cinemas by the mid sixties. By 1967 there were 41 kindergartens, 76 primary, 51 intermediate, 10 secondary and 17 other schools; 10 hospitals, 38 medical clinics, and the same number of dental centres; 11 child and mother welfare centres; 11 preventive medicine establishments and 195 school health clinics. There were 249 industrial establishments employing more than 20 workpeople each, and 32 hotels. Construction permits issued between 1966 and 1968 number slightly under 17,000, covering an area of about 7 million square yards. Scheduled agricultural areas diminished from 6·9 million square yards in 1967 to 4·8 million square yards in 1968. From 1953 to 1956, paved roads stretched for 268,000 yards. By 1968, that figure exceeded 1·8 million yards. And by this time, there were, in round figures, 90,000 private cars, 10,000 taxis, 28,000 vans and lorries, and 1,800 buses on the roads at an average density of about fifty-six vehicles to the mile, which puts Kuwait at the top of the world's traffic jam league, just alongside of Great Britain and Belgium.

The bustle of building activity and the turmoil of traffic-laden roads is not hard to imagine from such figures. Naturally enough, the graph of accidents and death on the roads has risen as sharply as the skyline. The new Kuwait has the best and the worst of the Western civilisation that it so assiduously set out to emulate. What of its Arab identity? What has remained to act as a visual and spiritual link with the past? Not much, but enough if planning plays a proper part in regimenting what has gone before by bringing it within the framework of a carefully disciplined architectural unity. The first auguries were not very good. Architects and civil engineers from virtually every country that could lay claim to a reputation in those spheres came in from Egypt, Morocco, Europe, Britain and America, armed with their own particular solutions and riding their own hobby horses. Kuwait was soon the possessor of more architectural styles and more bizarre planning arrangements to the acre than almost any place in the world, except perhaps Los Angeles and some English seaside resorts. By the turn of the sixties, a tour of the sprawling city and its ever-extending suburbs became an assault on the eyes. The speculators of the building world have left a costly trail of experiment and bad taste in their wake; as they always do unless their activities are controlled by a planning authority of stern and sensitive purpose. Architects undoubtedly used this as a rewarding stamping ground. Their perverse structures leave few of the rules of harmony unscathed. Verticals and horizontals are almost non-existent in domestic building. Cantilevered balconies project at strange angles. Walls are usually convex or concave and sloping inward or outward. There is little or no coherence about it all. Of the contrast between the indiscipline of this brave new world and the traditional architecture of Kuwait, Dr Saba Shiber has said:

"Its buildings were a closely-knit labyrinth that repelled heat and sandstorms . . . having thick walls, narrow apertures and properly located slits for ventilation. . . . Despite all the millions of dinars spent since the discovery of oil, one can safely say . . . that few are the new buildings which obey any of the laws of functionalism and organic design."

One thing they all forgot about in their hurry to get on with their own isolated contributions to this endlessly rich building boom, was old Kuwait. The entrepreneurs of the building world are notoriously insensitive. The 1920 wall was one of the first of the traditional structures to go. Many of the old houses by the sea front went soon after. They left the gates which allowed passage through the wall. A few of the early houses remained, including the old Political Agency which stayed in the possession of Mrs Dickson through the intervention of successive amirs. Another pleasant link with the past is provided by the palatial house built in 1890 by Shaikh Khazal of Muhammerah. In fact, it is two houses, now separated by a new road. The harem on one side has been occupied for many years by one of Kuwait's richest and most famous families, the Al Ghanims. The *diwaniyah* on the other side has been turned into Kuwait's only museum, housing wildlife, replica pearl divers and other pre-oil features, as well as some of the archaeological finds of the Falaika expedition. The Dasman Palace, former residence of Shaikh Ahmad al Jabir and now the home of the heir apparent, Shaikh Ahmad's son Jabir, still stands in splendid isolation.

It was a London firm of planning consultants that first tried to give cohesion to this mushroom development. But little of its plan was ever put into effect. A development board was set up in 1952 and this tried to co-ordinate the activities of a multiplicity of building schemes during the next critical decade. By the time a more powerful and wider-ranging State Planning Board had been formed in 1962, much of the damage had been done. At one moment a coastal pattern of domestic building was decided on. Then planning in depth, with inland townships growing up around existing habitations took its place. Between these South of France and Newcastle-upon-Tyne extremes of planning, the monoliths of neo-Levantine, neo-Egyptian, neo-anything went up. Here and there a string of domestic buildings would rise in isolation and remain in a no man's land for a year or two until somebody else came along and extended the string in another style. All the time, the wirescape spread through the towns and across the desert. Television aerials sprouted in profusion.

There were some notable exceptions to the indiscipline. The main luxury hotels were built in logically modern style and sensible locations. The new Seif Palace is an attractive piece of architecture. The university, originally the Shuwaikh Secondary School, was superbly planned and, in its greyish brown stone and almost Victorian manner, a distinctive building. The flourishing port and industrial centre of Shuwaikh and the great twentieth-century complex of Shuaiba are unrestrained in their tribute to industrialisation. It is in bringing

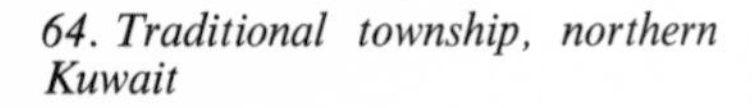

64. Traditional township, northern Kuwait

these diverse elements together in an unashamedly present-day scheme, while retaining – and even highlighting – surviving traditional features of the landscape, that the task of the planners meets its real challenge. This might well, in the beginning, have been a job for a Corbusier or a Lloyd-Wright, for an architectural dictator who could have used a wide open space and existing Arab traditions to create a new city of excitement and sensibility, and coherence. That was not to be, however. It was not until the task had become one of developing outward from what was already there, rather than of creating from scratch on virgin land, that another distinguished planner was called in. Professor Colin Buchanan, famous for his report on Britain's traffic problems, prepared a new plan for the Kuwait of tomorrow which was submitted to the planning board in draft form in 1970. There was, perhaps, a supreme relevance in appointing this consultant in a country whose greatest single problem in the future will almost certainly be mobility.

Master plans

The first Buchanan study was presented to the planning board and the municipality of Kuwait in March 1970. In its own words it was an "account of the proposed physical planning strategies for the long-

65. The new architecture. Present day street scene, Kuwait city

term development of the state of Kuwait and the urban areas". Like all grand strategies it had to make certain basic assumptions, on the strength of which the theoretical options were stated. The parameters were, of course, largely fixed before Professor Buchanan and his team got to Kuwait.

The London architects Minoprio, Spencely and Macfarlane had put up an earlier master plan during the 1950s. This embraced a "Corniche" scheme consisting of a series of four concentric ring roads with a super-block development between the rings and radials, each block covering an area of about 1,500 metres square. These were to provide mainly residential accommodation and had their own infrastructure of social and public utilities. Industrial development was to be confined to the Shuwaikh region, while sensible provision was made for open spaces and tree plantations.

Unfortunately building within the scheduled areas took on a staccato rhythm which was far from the smooth pattern of progress envisaged. A further study, undertaken by a distinguished international panel of architects, produced an interim plan. Broadly, this recommended a policy of containment within the outer ring road, together with coastal development in the south-east for both housing and industry. In 1967, the Town Planning Department set forth supplementary plans which extended the original scheme by the addition of two more ring roads, with radial roads extending to the sixth or outer ring. The Shuaiba industrial scheme south-east of Ahmadi was conceived at this time.

When Professor Buchanan came on the scene, a great deal of concrete and bitumen was irremovably in place. In order to prepare the ground for a long-term strategy, his team employed the kind of statistical analysis and projected planning criteria for which this architectural practice is renowned. In doing so, it threw an interesting light on the social and economic directions in which Kuwait was moving. Its assessment of population growth was subject to the fundamental proviso that the options are wide open to Kuwait as far as immigration is concerned. If the country wants a broader based economy calling for increasing professional skills and trained labour, it cannot avoid the corollary of growing immigration and an increase in the ratio of expatriates to Kuwaitis. Assuming a three per cent net immigration annually, the Buchanan report predicts a population of two million somewhere between 1985 and 1995. There was no need to attempt a more accurate forecast. As long as the period and the likely rate of growth were established, everything else could be phased to meet social needs arising from the population trend. By a fairly conservative extrapolation of the Buchanan graph, incidentally, the population will exceed 2½ million by the turn of the century.

The facts which follow from that forecast are of critical importance to all the social and economic problems confronting the country. By 1970, Kuwait's approach to economic policy was not unlike Britain's approach to membership of the European Economic Community. It could see no practical alternative to a course of action that it was reluctant to pursue. It wanted to broaden its manufacturing base and its overall productivity, even though the benefits might turn out to be marginal or even self-defeating. Some of the question marks are underlined by Buchanan. If population growth reaches its maximum by 1985, there will be need of 800,000 jobs. But, since the faster the growth the greater the proportion of immigrants, only 625,000 jobs would have to be found if the two million population figure was reached in 1997. The foreign worker is, of course, generally in the active age group in terms of employment and reproduction. The report suggests that over the next twenty years it will in fact be necessary to find some 560,000 new jobs, taking 1969 as a starting point, of which about three-quarters will be in the service sector.

Traffic is a problem which Kuwait shares with all economically advanced countries, though there is good reason to think that it might be worse here than elsewhere. Buchanan worked on estimates of eight and ten per cent growth rates in the car population and a saturation level of 1·8 vehicles per family (0·33 cars per head of population). The percentage growth rates make no appreciable difference in the long term but whether the higher or lower rate applies it will doubtless give rise to a greater deal of congestion in the meantime, to say nothing of an increase in the noise level, if that is possible in a community where horn blowing is a national pastime.

Land usage is another vital factor in any long-term plan. Buchanan adopted a figure of sixty people to the hectare (just under $2\frac{1}{2}$ acres)

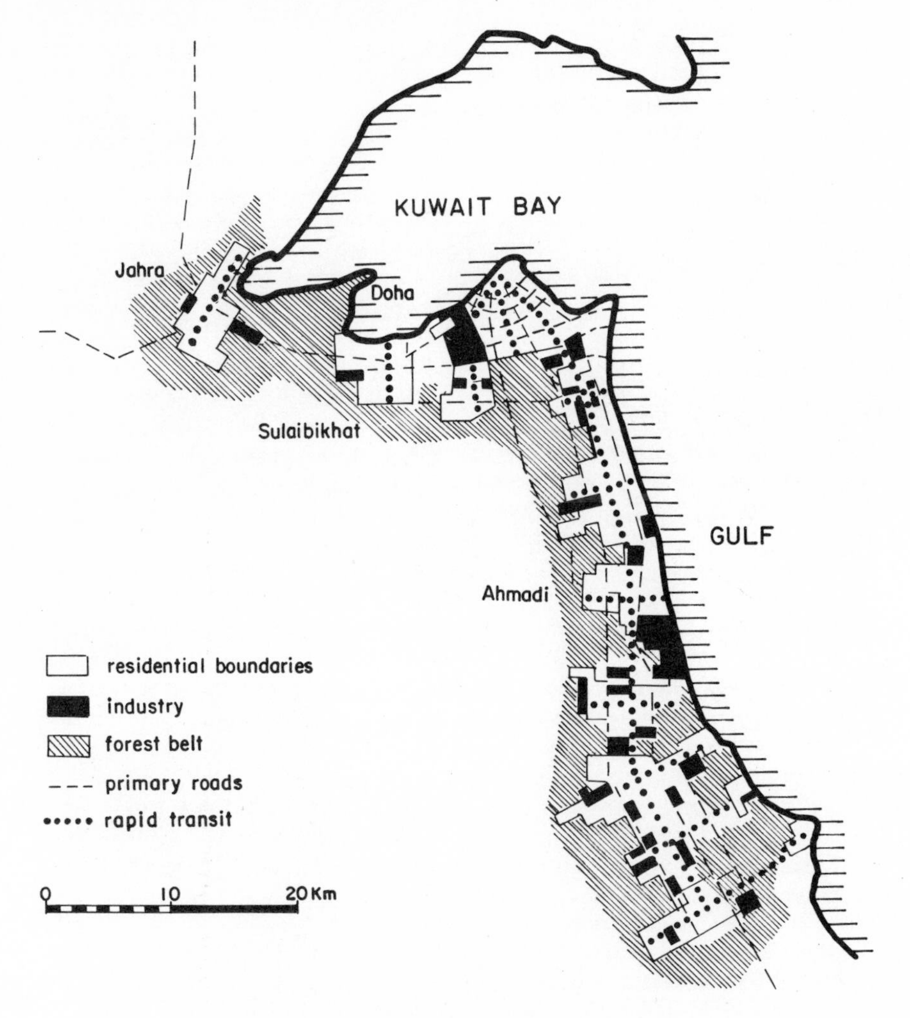

ix Possible realisation of the Buchanan scheme.

for low density residential areas, and 150 to the hectare for high density zones. Together with the land requirement for schools and essential amenities, the total land needed in the future for housing and associated use will be 20·2 hectares per 1,000 people. Thus, at the most critical postulate of two million people by 1985, the land covered by residential building, parkland and roads will be about 161 square miles. In the industrial sphere, the Kuwait Oil Company was occupying some 12,000 hectares over which it had exclusive surface rights by 1970. In addition, Aminoil occupied 3,000 hectares. The Buchanan team regard this as sufficient for expansion within the prescribed target of a 6 per cent yearly increase in oil output. There is, however, a potential need of space for the Kuwait Spanish Petroleum Company (KNPC and Hispanoil) and the Kuwait Shell Petroleum Company if and when they begin production from the relinquished KOC and offshore areas. Estimates for light and heavy industries, including extensions to the existing areas of Shuwaikh and Shuaiba, run to 1,924 and 142 hectares respectively. With an overall space allocation for roads of 10 per cent, and with allowances for university, stadia, district amenity centres and other requirements of 800 to 1,000 hectares, the total estimated land demand for urban development capable of absorbing a population of two million, would be in the region of 30,000 hectares, in addition to already committed sites; or 120 square miles.

66. One of Kuwait's several cinemas contributes to a panoply of architectural styles

The unquenchable Kuwaiti need of water – for household, industrial and irrigation use – is expected to generate an average demand of 600 million gallons a day within the next twenty years. Available resources from distillation plants, brackish wells, and underground sweet-water reservoirs, add up to 375 million gallons a day. Thus, a gap of 225 million gallons a day has to be bridged before 1990. There is only one plausible answer, the Shatt al Arab. In 1964, Kuwait concluded a treaty with Iraq permitting a take off of 120 million gallons a day. So far, this source has remained untapped, but it may well have to be used in future. Power, too, shows the same pattern of staggering increases in an already unusually high per capita demand. Between 1956 and 1959, the rate of growth of the power industry was nearly 28 per cent per annum. Buchanan assumes a per capita growth rate of 14 per cent until 1975, falling to 7 per cent from then on. The demand is expected to grow from 620 Mw in 1970 to 8,000 Mw in 1990. Here, Kuwait is fortunate in having virtually unlimited natural gas on its doorstep as a prime source of energy. But even this will be in critically short supply by 1975.

These, then, were the factual limits within which, given a policy of widespread industrial expansion, Professor Buchanan's team were able to design for the future. To plan an orderly, carefully phased, growth on foundations which were to a large extent laid.

Taking account of natural and artificial constraints such as climate, existing development, environment factors, and the network of oil installations cutting through the country, the new master plan combines an imaginative blueprint for living with what is inevitably a compromise between a desert community and a superimposed technology. Briefly, it designates, in what it calls a preferred strategy, areas of urban development stretching from the north-easterly limits of old Kuwait, eastward along the bay and then down the coast to a point opposite Ras al Jilaia; this urban belt makes full residential use of the southern coastline, with large recreational spaces at its northern and southern extremities, again making the most of the coastline.

The hinterland of the urban stretch is occupied by the oilfields, except for the region south-west of the bay around Jahra, which is set aside for the extension of its agricultural activity. Agricultural expansion is also planned for the larger part of Bubiyan island, while the northern oilfields are flanked by green zones stretching eastwards to the coast and joining up with the recreational area north of the bay. Failaka, with its excavated Dilmun and Greek sites, remains as a rural, tourist amenity. A large zone reaching from Jahra to the Saudi border, some eighty miles in overall depth, is a restricted military area. The airport, extended in size and facility, remains where it is presently situated between Shuwaikh and Ahmadi. The Ahmadi Ridge as it has always been called by the oil fraternity, presents difficulties for the planners since much of it is occupied by the installations and the massive tank farms of the Kuwait Oil Company. The Zor escarpment across the bay is also partly inaccessible because of military claims on it. However, KOC in 1971 expressed a willingness to relinquish some parts of the ridge and if the military authorities, who reserve part of the Zor escarpment as a training ground, could be induced to do the same, the architects see these as major features of their amenity plans; features which could be made into high points of great visual interest. While recognising the need to expand the existing industrial zones of Shuwaikh and Shuaiba, Buchanan proposes eleven new light industrial areas. Three of these are to the west of Kuwait city, three in the eastern coastal strip, and five in a large satellite development south of Ahmadi, coincident with the southward residential spread. A new site for extra power and water plant would be located north of Fantas.

67. Greenery, elegant streetlights and well-conceived modern building; another aspect of Kuwait's architecture

These new development areas lie mainly to the south of existing urban districts. They divide into five regions; the coastal strip between the fifth ring road and committed developments north of a line drawn from Ahmadi to Fahaheel, a new satellite south of Ahmadi, an area south of the coastal town of Sulaibikhat, and extensions to Jahra and Ahmadi. Between them these new and expanded development regions would accommodate 1,105,000 people. In addition, Hawalli and Salimiya would be expanded to cater for populations of up to 140,000 each. As for Kuwait itself, a residential capacity of 100,000 is envisaged with jobs for 110,000, thus it is hoped avoiding a large-scale commuter problem. Motorway-Expressway communication between the major southern satellite and Kuwait city is, of course, catered for.

Finally, the scheme provides for an attractive tree planting operation. In fact, for a possible major forest belt, wrapping round the southern satellite, skirting Ahmadi and Sulaibikhat, and finishing up towards the coast, north-east of Jahra. It is a well conceived scheme, and whatever parts of the plan are adopted it will cost a lot of money – over twenty years, something of the order of KD 5,000 million, excluding machinery and equipment.

In the last resort, its success will depend on attention to detail, as builders have discovered but seldom appreciated since the times of the ancient civilisations to which the place that is now Kuwait was once witness. Perhaps in the age of concrete and steel, a new order will be restored out of the existing complexities of style and experiment. Whatever the Buchanan proposals achieve or fail to achieve, it is to be hoped that they will be instrumental in preserving the little that, by the 1970s, remained of old Kuwait.

If there is an implied impertinence in telling countries other than one's own how they should manage their affairs, better excuse may be found for reversing the process by taking a glance at the follies and frailties to which affluence has led in the West. If Kuwait were to pause and examine American and European society in the wake of more than a century of intensive industrialisation, some useful lessons might be learned, and salutary experience gained. In a world where cities grow ever thicker on the ground and higher above the skyline, there is never enough accommodation for the millions who seek after the ephemeral promise of a better life. In societies dedicated to the car and fast transport, roads are never long enough or wide enough for five minutes at a time. It is a lucky citizen who can park his automobile outside his own house without threat of a fine, or putting money into a parking meter for the privilege. At the end of the day, the highest priority of such societies is the provision of hospital beds; and one in three of those is occupied by a mental patient. There seems to be an inexorable rule that progress in the quantitative sense sooner or later gives rise to qualitative change – usually for the worse. It may be too late for America and Europe to learn the lesson, or at any rate to do anything about it. Countries like Kuwait, and there are few enough of them, with resources and space sufficient to use or conserve with a thought for the future, can afford to stand back and take stock, before they come too close to the point of no return.

Bibliography

Anderson, J. R. L, *East of Suez*, Hodder and Stoughton, 1969.
ARAMCO, *Handbook*, Arabian American Oil Company, 1960.

Bibby, Geoffrey, *Looking for Dilmun*, Collins, 1970.
British Petroleum, "BP: Our Industry", Private circulation.
Buchanan, "Studies for National Physical Plan and Master Plan for Urban Areas", Colin Buchanan and Partners, 1970.

Calverley, Eleanor, *My Arabian Nights and Days*, Thomas Y. Crowell Co, New York 1958. Out of print.

Daniel, Glyn, *The Idea of Prehistory*.
Dickson, H. R. P., *Kuwait and Her Neighbours*, George Allen and Unwin, 1956.
— *The Arab of the Desert*, George Allen and Unwin, 1949.
Dickson, Violet, *Forty Years in Kuwait*, George Allen and Unwin, 1970.

Fox, A. F., "Symposium Sobre Yacimuntoo de Petrolo Y Gas", 20th Mexico Int. Icol. Congress 1956, Vol. 2.
Freeth, Zahra, *Kuwait was My Home*, George Allen and Unwin, 1956.
—*A New Look at Kuwait*, George Allen and Unwin, 1972.

Hamilton, Charles W., *Americans and Oil in the Middle East*, Gulf Publishing Company, Texas 1962.
Hansen, Thorkild, *Arabia Felix: The Danish Expedition 1761–67*. Collins, 1964.
Harrison, Paul, *The Arab at Home*, 1923. Out of print.
— *Doctor in Arabia*, Out of print.
Holden, David, *Farewell to Arabia*, Faber and Faber, 1966.

Longhurst, Henry, *Adventure in Oil: the Story of British Petroleum*, Sidgwick and Jackson, 1959.
Longrigg, Stephen Hemsley, *Oil in the Middle East*, Oxford University Press, 1968.

Mallakh, Regaei el, *Economic Development and Regional Co-operation: Kuwait*, University of Chicago Press, 1968.
Marlowe, John, *The Persian Gulf in the Twentieth Century*, Cresset Press, 1962.
Mikdashi, Zuhayr, *A Financial Analysis of Middle East Oil Concession 1901–1965*, Frederick A. Praeger, New York, 1966.
Milton, D. I., *Geology of the Arabian Peninsula*, Professional paper, US Government Printing Office, Washington, 1967.
Mineau, Wayne, *The Go Devils*, Cassell, 1958.
Ministry of Guidance and Information, *Kuwait Today*, Quality Publishers, Nairobi, 1963.
— *Archaeological Investigations in Island of Failaka 1958–64*, Kuwait Government Press.
Monroe, Elizabeth, *Britain's Moment in the Middle East*, Chatto and Windus, 1964.
Morkholm, Otto, "Coins Found on Failaka", Paper for Department of Antiquity and Museums, Kuwait Government Press.

Nutting, Anthony, *The Arabs*, Hollis and Carter, 1964.

Odell, Peter R., *Oil and World Power*, Penguin, 1970.
Owen, R. M. S. and Nasr, Sami N., "Stratigraphy of the Kuwait-Basra Area", Symposium of American Association Petroleum Geologists, 1958.

Philby, H. St John, *Arabian Highlands*, Cornell University Press.
— *Arabian Days*, Robert Hale, 1948.
Planning Board, *Kuwait Economic Survey 1968-69*, Kuwait Government Press.
— *The First Five Year Plan 1967-72*, Kuwait Government Press.

Raunkiaer, Barclay, *Through Wahhabiland on Camelback*, (introduced by Gerald de Gaury), Routledge and Kegan Paul, 1969.

Shiber, George S., *The Kuwait Urbanisation*, Kuwait Government Press, 1964.
Shushtery, A. M. A., *Outlines of Islamic Culture*, Bangalore Press, 1938.

Tugendhat, Christopher, *The Biggest Business*, Eyre and Spottiswoode, 1968.

Van Ess, John, *Meet the Arab*, Museum Press, 1947. Out of print.

Winder, R. Bayly, *Saudi Arabia in the 19th Century*, Macmillan, 1965.
Wilson, Sir A. T., *The Persian Gulf*, London, 1928.

Index